DRIVES AND SEALS

DRIVES AND SEALS

Edited by

M. J. NEALE
OBE, BSc(Eng), DIC, FCGI, WhSch, FEng, FIMechE

A TRIBOLOGY HANDBOOK

Published by:
Society of Automotive Engineers, Inc.
400 Commonwealth drive
Warrendale, PA 15096 - 001

Butterworth-Heinemann Ltd
Linacre House, Jordan Hill, Oxford OX2 8DP

A member of the Reed Elsevier group

OXFORD   LONDON   BOSTON
MUNICH   NEW DELHI   SINGAPORE   SYDNEY
TOKYO   TORONTO   WELLINGTON

First published 1994

© The editor and contributors 1994

All rights reserved. No part of this publication
may be reproduced in any material form (including
photocopying or storing in any medium by electronic
means and whether or not transiently or incidentally
to some other use of this publication) without the
written permission of the copyright holder except in
accordance with the provisions of the Copyright,
Designs and Patents Act 1988 or under the terms of a
licence issued by the Copyright Licensing Agency Ltd,
90 Tottenham Court Road, London, England W1P 9HE.
Applications for the copyright holder's written permission
to reproduce any part of this publication should be addressed
to the publishers

Library of Congress Cataloguing in Publication Data
A catalog record for this book is available from the
Library of Congress

ISBN 1 56091 452 1

SAE order number: R - 138

Printed and bound in Great Britain

Contents

Editor's Preface vii
List of Contributors ix

Rotary drives

1. Belt drives 1
2. Roller chain drives 11
3. Gears 17
4. Flexible couplings 25
5. Self-synchronising clutches 32
6. One way clutches 35
7. Friction clutches 37
8. Brakes 46

Linear drives

9. Screws 54
10. Cams and followers 57
11. Wheels rails and tyres 63
12. Capstans and drums 70
13. Wire ropes 72
14. Control cables 74
15. Damping devices 76
16. Pistons 79
17. Piston rings 87
18. Cylinders and liners 94

Seals

19. Selection of seals 97
20. Sealing against dirt and dust 105
21. Oil flinger rings and drain grooves 109
22. Labyrinths, brush seals and throttling bushes 111
23. Lip seals 117
24. Mechanical seals 122
25. Packed glands 132
26. Mechanical piston rod packings 136
27. Soft piston seals 140

Index 144

Editor's Preface

This handbook gives practical guidance on drives and seals in a form intended to provide easy and rapid reference. It is based on material published in the first edition of the *Tribology Handbook* and has been updated and matched to international requirements.

Each section has been written by an author who is expert in the field and who in addition to understanding the related basic principles, also has extensive practical experience in his subject area.

The individual contributors are listed and the editor gratefully acknowledges their assistance and that of all other people who have helped him in the checking and compilation of this revised volume.

Michael Neale
Neale Consulting Engineers Ltd
Farnham 1993

Contributors

Section	Author
Belt drives	T. H. C. Childs BA, MA, PhD, CEng, FIMechE, MInstP
Roller chain drives	M. C. Christmas BSc, CEng, MIMechE, MIMgt
Gears	A. Stokes
Flexible couplings	M. J. Neale OBE, BSc(Eng), DIC, FCGI, WhSch, FEng, FIMechE
Self-synchronising clutches	J. Neeves BA(Eng)
One-way clutches	T. A. Polak MA, CEng, MIMechE
Friction clutches	T. P. Newcomb DSc, CEng, FIMechE, FInstP, CPhys R. T. Spurr DSc, PhD, DIC, FInstP, CPhys H. C. Town CEng, FIMechE, FIProdE
Brakes	T. P. Newcomb DSc, CEng, FIMechE, FInstP, CPhys R. T. Spurr DSc, PhD, DIC, FInstP, CPhys
Screws	M. J. Neale OBE, BSc(Eng), DIC, FCGI, WhSch, FEng, FIMechE
Cams and followers	T. A. Polak MA, CEng, MIMechE, C. A. Beard CEng, FIMechE, AFRAeS
Wheels rails and tyres	W. H. Wilson BSc(Eng), CEng, MIMechE
Capstans and drums	C. M. Taylor BSc(Eng) MSc. PhD, DEng, CEng, FIMechE
Wire ropes	D. M. Sharp
Control cables	G. Hawtree C. Derry
Damping devices	J. L. Koffman DiPlIng, CEng, FIMechE
Pistons	B. L. Ruddy BSc, PhD, CEng, MIMechE G. Longfoot CEng, MIMechE
Piston rings	R. Munro BSc, PhD, CEng, MIMechE B. L. Ruddy, BSc, PhD, CEng, MIMechE D. C. Austin
Cylinders and liners	E. J. Murray BSc(Eng), CEng, MIMechE N. Tommis AIM, MIEI, AIMF
Selection of seals	B. S. Nau BSc, PhD, ARCS, CEng, FIMechE, MemASME
Sealing against dirt and dust	W. H. Barnard BSc(Lond), CEng, MIMechE
Oil flinger rings and drain grooves	A. B. Duncan BSc, CEng, FIMechE
Labyrinths, brush seals and throttling bushes	B. S. Nau BSc, PhD, ARCS, CEng, FIMechE, MemASME
Lip seals	E. T. Jagger BSc(Eng), PhD, CEng, FIMechE
Mechanical seals	A. Lymer BSc(Eng), CEng, FIMechE, W. H. Wilson BSc(Eng), CEng, MIMechE
Packed glands	R. Eason CEng, MIMechE
Mechanical piston rod packings	J. D. Summers-Smith BSc, PhD, CEng, FIMechE R. S. Wilson MA
Soft piston seals	R. T. Lawrence MIED

Belt drives 1

BELT DRIVE DESIGN

(a) Basic 2-shaft drive

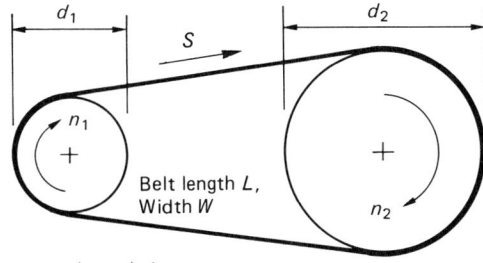

d in mm, n in rev/min
Belt speed $S = nd/19100$ m/s

(b) Idler added to increase arc of contact or control belt vibration

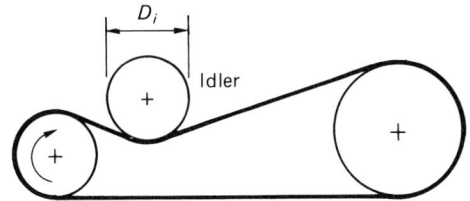

(c) Serpentine layout to drive more than one shaft or reverse direction of rotation

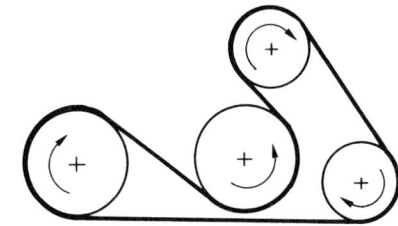

(d) Out-of-plane layout

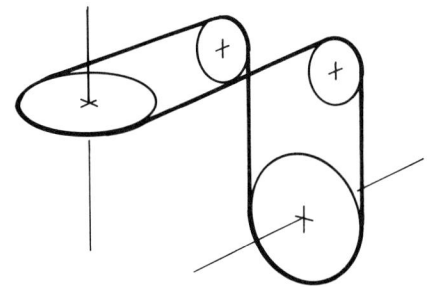

Figure 1.1 Some belt drive layouts

Belt selection, pulley size and shaft load calculation procedures for drives with two shafts

- Specify rated power and pulley rev/min requirements
- Calculate design power from rated power × service factors
- Select belt section from design power rating charts, on basis of design power and rev/min of fastest pulley. Choose pulley diameters greater than minimum recommended.
- Select belt width (or number of belts, for Vee belts) from belt capacity tables
- Calculate belt tensions and shaft loading to limit slip (or to ensure good meshing of synchronous belts)

A typical two shaft belt arrangement is as in Figure 1.1(a).

Some guides recommend selection on basis of rated power. This can result in smaller diameter pulleys and lighter section belts but a wider drive.

Pulley diameter and belt section selection is a compromise between smaller diameter pulleys for compactness, but limited by overloading of the tension members in the belts, and larger pulleys giving lower bearing loads but requiring more space. Larger pulley diameters may always be chosen, subject only to allowable belt speeds, and the designer is advised to check the effect of varying diameter by trial calculations.

Belt width and tension selection are closely related and depend on arc of contact as well as power and rev/min.

1 Belt drives

BELT TYPES AND MATERIALS SELECTION

Standard Vee, Vee-ribbed and synchronous (timing) belts have a high modulus wound-cord tension member (glass fibre, polymeric or steel members are used according to application), a rubber carcass and a woven fabric backing cover. The drive surface of synchronous belts is strengthened by a woven fabric cover, that of Vee-belts may be covered or uncovered (raw), while Vee-ribbed belts are generally raw. The flat belts considered here have polyamide strip tension members and either rubber or leather drive surfaces. This handbook does not consider all variations of belt section and materials that exist. Users should consult catalogues for the full range of constructions and materials.

The operating temperature range of belts is typically $-20°C$ to $+70°C$, but materials may be formulated for both lower ($-40°C$) or higher ($+120°C$) temperatures. The static conductivity of belts is regulated by standards (e.g. ISO 1813, ISO 284, ISO 9563). Belts should be shielded from oil but most types are resistant to small amounts of contamination.

Power transmission efficiencies of 95% to 98% in steady operating conditions are achievable with all well maintained belt drives.

Table 1.1 Selection of the type of belt

WEDGE AND CLASSICAL VEE BELTS

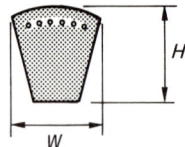 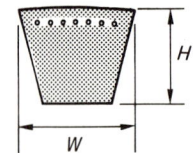

Further information
BS 3790, ISO 1081, DIN 2211, RMAIP 20
ISO 4184, DIN 7753, RMAIP 23
and manufacturer's catalogues

These are the standard choice for large power transmissions where slip in the event of shock loading is needed as overload protection. Wedge belts with a larger ratio of H to W than classical Vee belts give more compact drives but cannot be used in layouts requiring reverse bending of the belt. Stock pulley sizes allow speed ratios up to 8:1. Recommended maximum belt speeds are 30 to 40 m/s.

VEE-RIBBED BELTS

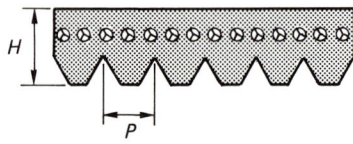

Further information
ISO 8372, DIN 7876, RMAIP 26
ISO 9982 and manufacturer's catalogues

These have been developed to combine the grip of Vee belts and almost the flexibility of flat belts and find application where space is confined (smaller diameter pulleys) or where some serpentine layout capability is needed. Stock pulley sizes allow higher speed ratios than for Vee belts, up to 25:1 depending on belt section. Recommended maximum belt speeds are 35 to 45 m/s.

FLAT BELTS

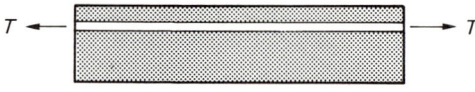

The cross-sections of flat belts are not controlled by Standards. Flat belts are rated by $k_{1\%}$, the load per unit width to stretch the belt 1%.
Further information from manufacturer's catalogues

The ease of joining polyamide strip tension member belts enables them to be made of virtually any length; their flexibility makes them suitable for highly serpentine or out of plane layouts. At low belt speeds, drives are less compact and bearing loads are higher than for other belt types, but as speed increases above 20 m/s, flat belts come into their own. Speeds up to 70 m/s are possible.

SYNCHRONOUS BELTS

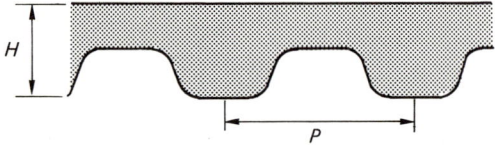

Further information
BS 4548, ISO 5296, DIN 7721
and manufacturer's catalogues

Synchronous belts are now developed to a similar power capacity as Vee belts. They are clearly essential if synchronous motion is needed but can suffer from tooth failure in conditions of extreme shock loading. The earlier developed trapezoidal toothed belt (illustrated) has now been displaced by curvilinear toothed belts in new drive designs. Typical maximum speed ratios are 10:1 and belt speeds are up to 60 m/s.

Belt drives 1

SERVICE FACTORS (SF) FOR VARIOUS APPLICATIONS

Design power = SF × Rated power.

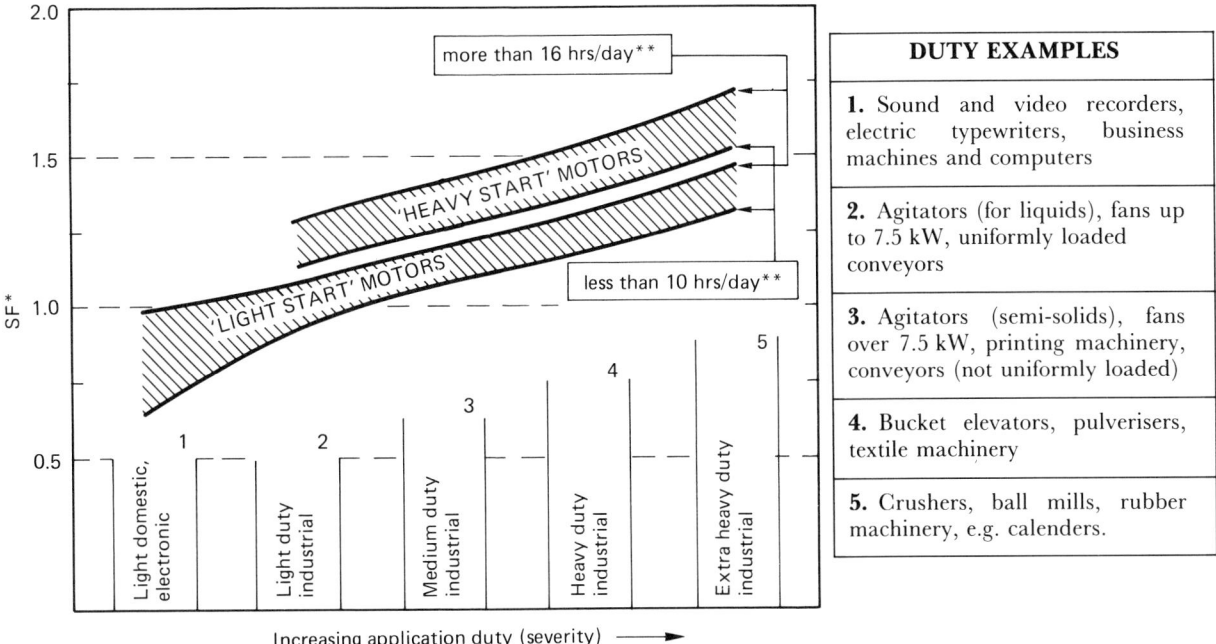

* Some design procedures include arc of contact factors in SF. This is not followed here. SF may be increased by up to 0.6 for speed-up drives and if the belt layout requires reverse bending. SF also depends slightly on belt length and other factors: this detail is not considered here.
** It is not customary to include operating time/day in SF for flat belt drives. A mid-range value may be chosen for these.

Figure 1.2 Service factors for various duties

BELT SELECTION FROM DESIGN POWER RATING CHARTS

The belt section is selected on basis of the region which contains the intersection of design kW with fastest pulley rev/min. Regions can overlap – more than one belt section may be a possibility.

Drives may be designed with any pulley diameters > d_{min} for that section. It is usually desirable to choose the largest diameter pulleys for which space is available, subject to the maximum recommended belt speed not being exceeded.

The labels $d_1 \times w_1$, $d_2 \times w_2$ give indications of likely combinations of smallest pulley diameter and belt width for 180° arc of contact drives.

The design power rating charts on the following pages are all consistent with the service factors SF above. They differ from some catalogues' charts which use differently based SF values.

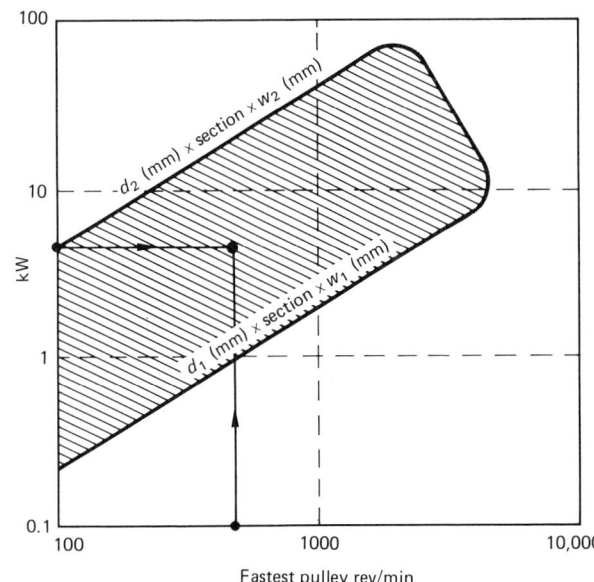

Figure 1.3 The use of power rating charts

1 Belt drives

DESIGN POWER RATINGS FOR VARIOUS TYPES OF BELT

Wedge belts

Section	SPZ	SPA	SPB	SPC	8V****
W mm	9.5	13	16	22	26
H mm	8	10	14	18	23
$\beta°$	40	40	40	40	40
d_{min} mm*	67	100	160	224	335
(d_{min} mm	56	80	112)		
D_i min	reverse bending is not allowed				
L_{min} mm**	525	750	1260	2000	2520
L_{max} mm**	3560	4500	9010	12500	11410
W_{max} mm***	72	90	152	306	343

* The d values here and in the power rating chart are for covered belts. For raw-edge moulded-cog wedge belts smaller sizes are appropriate, as shown by (bracketed values).
** Typical stock values (also covered by Standards); other lengths to special order.
*** Max. number of belts per pulley × width of one pulley groove l, (Table 1.2).
**** Rubber manufacturers of America (RMA) standard.

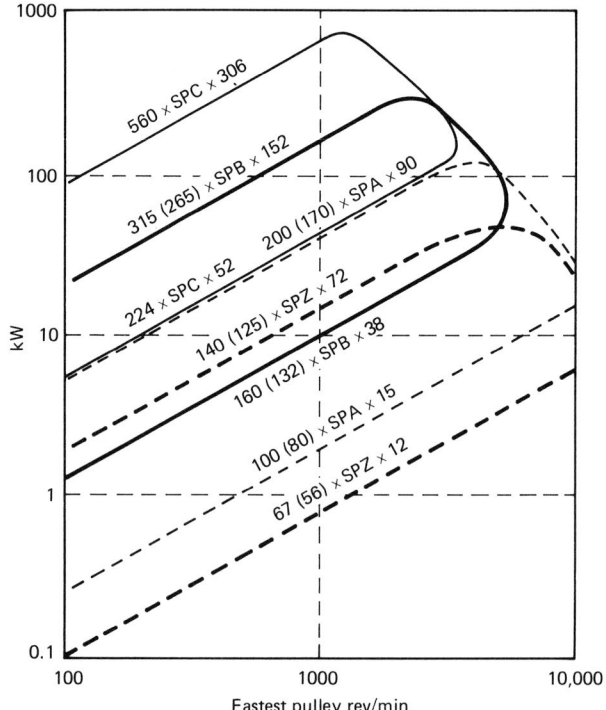

Figure 1.4 Power rating of wedge belts

Classical vee belts

Section	Z	A	B	C	D
W mm	10	13	17	22	32
H mm	6	8	11	14	19
$\beta°$	40	40	40	40	40
d_{min} mm	50	75	125	200	355
D_i min	> Smallest loaded pulley diameter				
L_{min} mm**	270	415	613	920	2570
L_{max} mm**	1540	5510	12000	10700	15200
W_{max} mm***	72	90	152	306	444

, * As above.
β is the total included angle of the belt profile.

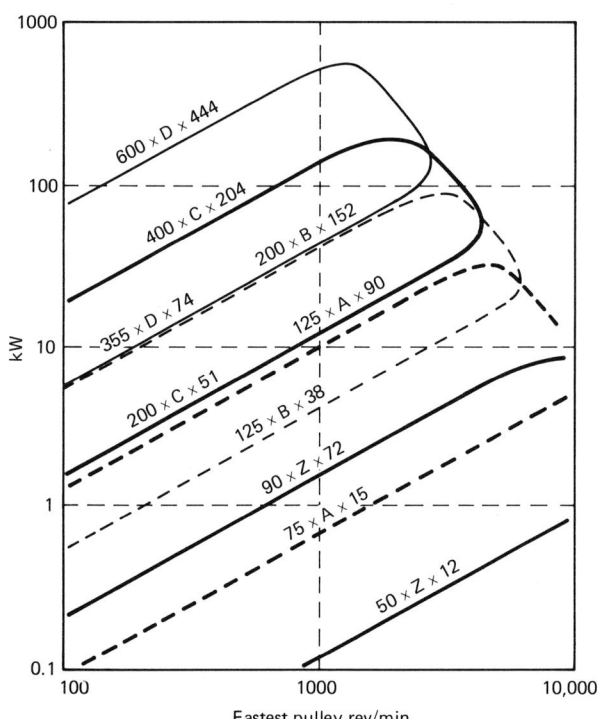

Figure 1.5 Power rating of classical Vee belts

Belt drives 1

Vee-ribbed belts

Section	J	K	L	M
P mm	2.3	3.6	4.7	9.4
H mm	Varies with manufacturer.			
$\beta°$	40	40	40	40
d_{min} mm*	20	38	76	180
$D_{i,min}$ mm*	32	75	115	267
L_{min} mm*	450	–**	1250	2250
L_{max} mm*	2450	–**	5385	12217
W_{max} mm***	46	72	94	188

* Varies with manufacturer.
** K is designed for automotive use but can be used for general machinery; some manufacturers also have an H-section, with $P = 1.6$ mm, and $d_{min} = 12.7$ mm.
*** W_{max} is 20 ribs for stock belts.

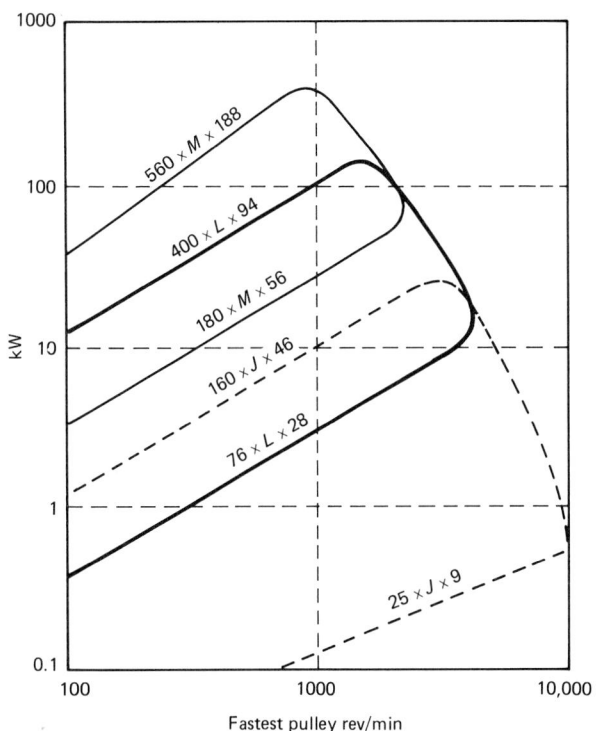

Figure 1.6 Power rating of Vee-ribbed belts

Polyamide core flat belts

Section*	a	b	c	d
$k_{1\%}$ N/mm**	3	7	15	25
d_{min} mm***	45	90	180	360
$D_{i,\,min}$ mm***	45	90	180	360
L_{min} mm	there is virtually no limit on belt length			
L_{max} mm				
W_{max} mm	< smallest pulley diameter			

* a, b, c, d have no Standards significance. The letters refer to Figure 1.7 and Figure 1.12.
** See Table 1.1. The $k_{1\%}$ values given here are a selection from those available. Values of 5, 10, 20, 33 and 40 N/mm may be obtained from a range of manufacturers.
*** Minimum pulley diameters vary with belt thickness, H. H can take a wide range of values for any $k_{1\%}$, as the ratio of friction material to tension member thickness varies with application. d_{min} and $D_{i,\,min}$ values given here are smallest values.

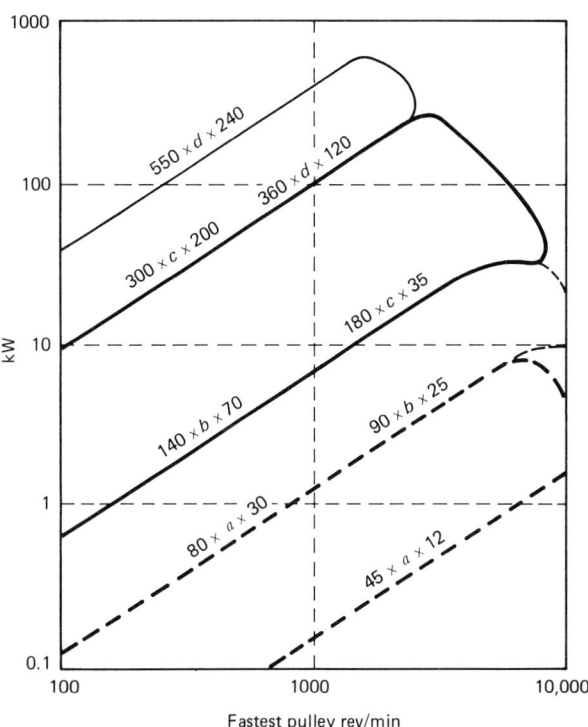

Figure 1.7 Power rating of flat belts with polyamide cores

1 Belt drives

Curvilinear tooth belts

The pitches below are de facto but not formal standards

P mm	3	5	8	14	20
H mm*	2.4	3.8	6	10	13.2
d_{min} mm	9.5	22	56	125	216
D_i mm	> smallest loaded pulley diameter				
L_{min} mm**	129	245	320	966	
L_{max} mm**	1863	2525	4400	6860	
W_{max} mm	< smallest pulley diameter				

* Varies slightly with manufacturer.
** Stock lengths, varies with manufacturer, other lengths to special order.

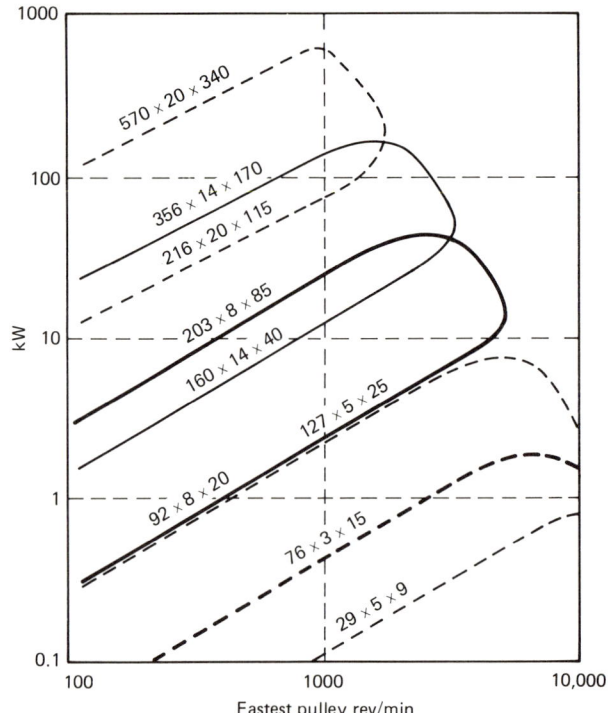

Figure 1.8 Power rating of curvilinear tooth belts

Trapezoidal tooth belts

Section	XL	L	H	XH	XXH
P mm	5.08	9.52	12.7	22.22	31.75
H mm	2.3	3.6	4.3	11.2	15.7
d_{min} mm	16	36	65	156	222
D_i, mm	> smallest loaded pulley diameter				
L_{min} mm**	152	314	609	1289	1778
L_{max} mm**	685	1524	4318	44445	4572
W_{max} mm	< smallest pulley diameter				

** As above.
The d values in the Figures 1.8 and 1.9 are nominal. Actual values give an integral number of teeth per pulley.

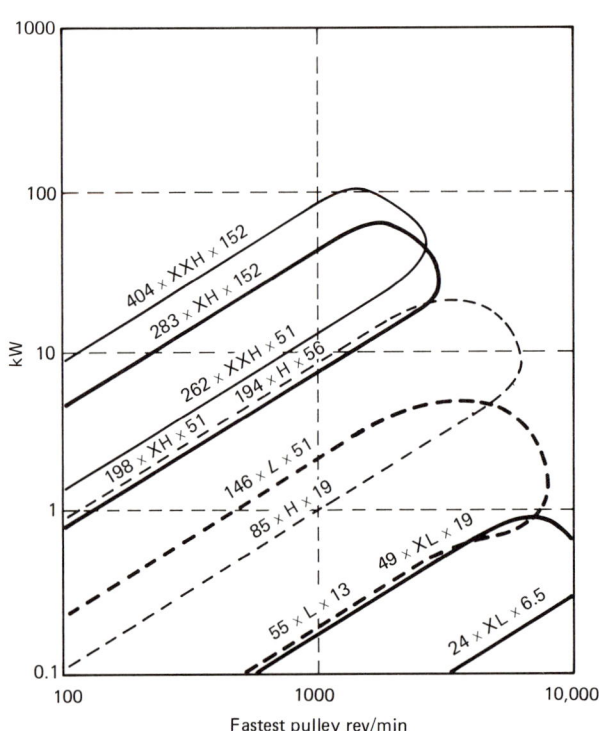

Figure 1.9 Power rating of trapezoidal tooth belts

Belt drives 1

BELT TENSIONS

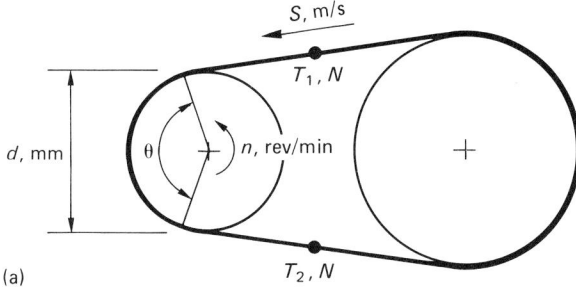

Tension sum $T_1 + T_2$ must be large enough to limit slip or, for synchronous belts, poor meshing.

$$T_1 + T_2 \geq (T_1 - T_2)/\lambda \qquad (2)$$

where traction coefficient λ varies with belt type and arc of contact.

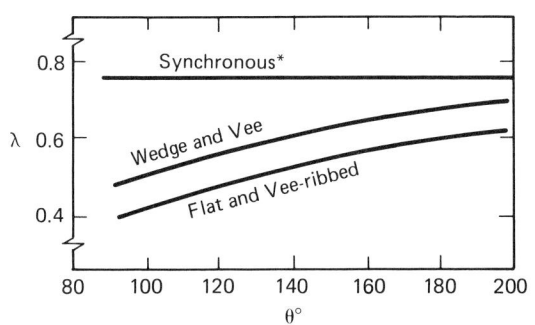

*provided more than 6 teeth are in mesh

Figure 1.10 Traction coefficient at various arcs of contact

Tension difference $T_1 - T_2$ arises from torque or power transmission

$$(T_1 - T_2).d = 19.1 \times 10^6 \left(\frac{kW}{n}\right) \qquad (1)$$

or $(T_1 - T_2) = 10^3 \left(\dfrac{kW}{S}\right)$

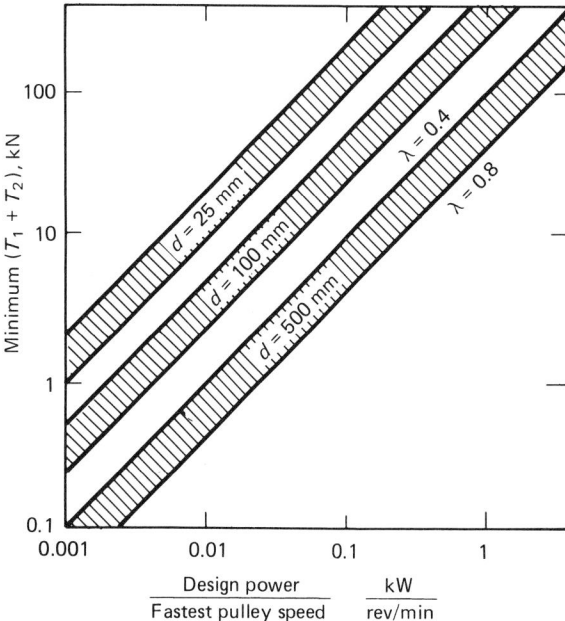

Figure 1.11 Minimum $T_1 + T_2$ at various operating conditions

BELT WIDTH

Belt width must be large enough to support tension. From (1) and (2) above, tension increases as $kW/(nd\lambda)$. Design guides tabulate allowable $kW/(nd)$ per belt or mm belt width, for $\theta = 180°$ arc of contact and varying n and d. Figure 1.12 gives values of F^*, from which such tables can be created.

To use the chart below, use θ for the smallest pulley to estimate λ_θ. Calculate

$$F = 19.1 \times 10^6 \left(\frac{kW}{nd}\right)\left(\frac{\lambda_{180}}{\lambda_\theta}\right)$$

Belt width = F/F^*, mm.

Bending a belt round a pulley increases tension member strain. Thus F^* reduces with reducing pulley diameter. Values below are mean values. F^* also reduces with increasing belt speed (see Figure 1.13)

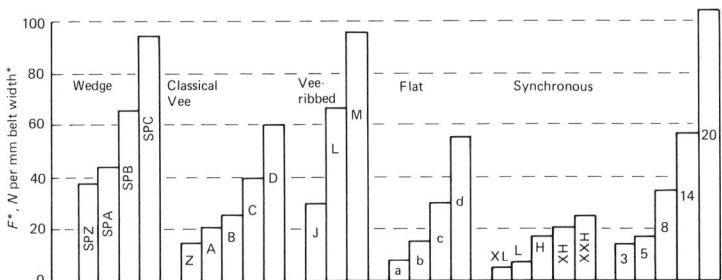

For wedge, Vee and Vee-ribbed belts, it is more usual to record F^* as N per belt or per rib. This can be derived from the above by multiplying by belt or rib width (mm). The data for wedge belts are for covered types: for raw-edge moulded-cog wedge belts of SPZ, SPA and SPB Section, F^* should be increased by 25–30%.

Figure 1.12 F^* the allowable belt tension difference per unit width for various types of belt

7

1 Belt drives

SHAFT LOADING

Static or installed shaft load

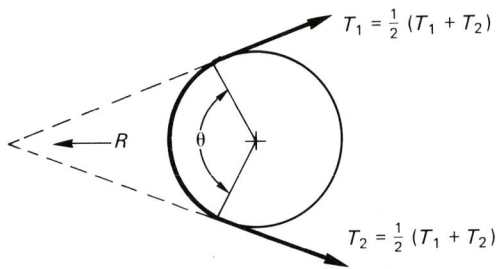

$R = (T_1 + T_2) \sin (\theta/2)$

Except for synchronous belts, $(T_1 + T_2)$ is obtained from Figure 1.11. For synchronous belts, $(T_1 + T_2)$ is 50%–90% of value from Figure 1.11 increasing with severity of shock loading: $(T_1 + T_2)$ rises to 100% during power transmission.

Dynamic or running shaft load

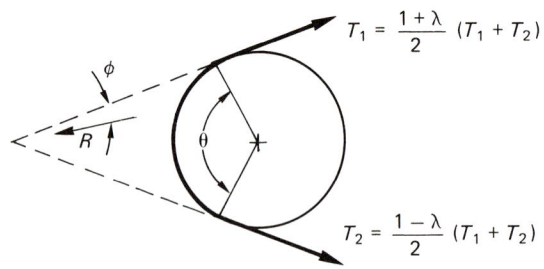

$R = (T_1 + T_2) \left\{ \dfrac{(1 + \lambda^2)}{2} - (1 - \lambda^2) \dfrac{\cos \theta}{2} \right\}^{1/2}$

$\sin \phi = \sin \theta \ (R/T_2)$

SPEED EFFECTS

Allowable kW/(nd) per belt or belt width reduces as S increases above 1 m/s, for Vee, Vee-ribbed and synchronous belts for two reasons. Up to $\simeq$ 20 m/s a faster speed simply means the belt is used more in a given time. Derating maintains its absolute life time. Over 20 m/s centrifugal loading becomes more significant. For both reasons belt width and shaft loadings are increased relative to the values obtainable from Figure 1.11 and Figure 1.12 by a speed dependent factor f.

Flat belts are, in practice, only de-rated for centrifugal loading.

ISO 5292 provides a model for the power rating of Vee-belts with respect to both speed and pulley diameter. It could be applied to all belt types.

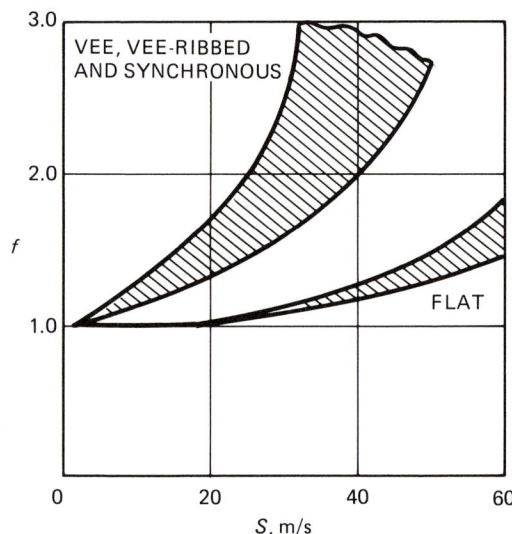

Figure 1.13 The derating in belt performance at higher speeds

Pulley materials for high belt speeds

The maximum safe surface speed of cast iron pulleys is 40 m/s. Steel of 430 MPa tensile strength may be used up to 50 m/s and aluminium alloys of 180 MPa tensile strength up to 60 m/s. Aluminium alloys are not recommended for uncovered rubber drive faces because of wear/abrasion problems with aluminium oxide. For operation up to 70 m/s special designs using high strength steel or aluminium alloys are required. Plastics are commonly used for low-speed, low power applications.

Belt drives

PULLEY DESIGN

Table 1.2 Wedge, Vee and Vee-ribbed pulley groove dimensions (mm)

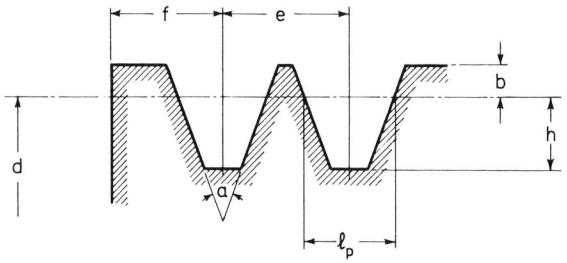

	l_p	b_{min}	h_{min}	e	f
Z	8.5	2.08	7.1	12	8
A	11	3.3	8.7	15	12.5
B	14	4.2	10.8	19	17
C	19	5.7	14.3	25.5	24
D	27	8.1	19.9	37	29
SPZ	8.5	2	9	12	8
SPA	11	2.8	11	15	10
SPB	14	3.5	14	19	12.5
SPC	19	4.8	19	25.5	17
J		−0.38*	2.18	2.34	1.8
L		−0.74	5.39	4.70	3.3
M		−1.47	11.05	9.40	6.4

* For Vee-ribbed belts the pitch line lies outside the pulley, and the groove tips and roots are rounded.

Table 1.3 Vee and wedge groove angles α related to pitch diameter (mm)

	32°	34°	36°	38°	40°
Z SPZ		<80		>80	
A SPA		<118		>118	
B SPB		<190		>190	
C SPC		>315		>315	
D			<500	>500	
J					>20
L					>76
M					>180

Users should not attempt to manufacture their own synchronous pulleys.

Surface finish

Standard recommendations for pulley drive face surface finish vary from 1.6 to 6.3 μm R_a (R_z from 6.3 to 25 μm), with 1.6 to 4 μm R_a being most common.

Further information
ISO 254, ISO R468, DIN 111, DIN 4768
and manufacturers' catalogues

Flat belt pulley crowning requirements

Both pulleys should be crowned. If one pulley is flat the crown of the other should be increased by 50%. A similar increase should also be applied when the shafts are vertical or when the centre distance is short.

Further information ISO 100, DIN 111

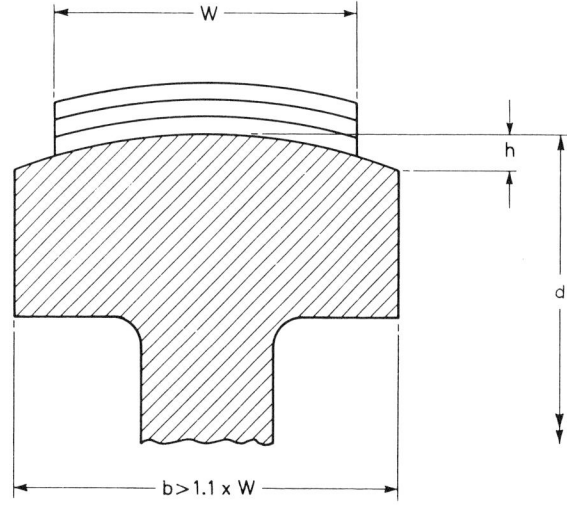

Avoid flanges if possible. If used, the corner must be undercut and b ⩾ 1.2W.

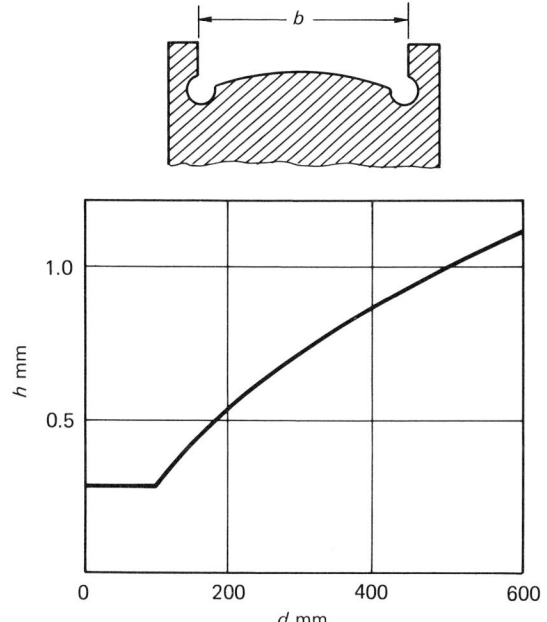

Figure 1.14 The amount of crowning required for flat belt pulleys

1 Belt drives

FITTING AND TAKE-UP ALLOWANCES

Fitting allowance (mm)

	Type	mm
V-belt	A	19–25
	B	25–40
	C	38–50
	D	50–75
Wedge belt	SPZ	12–20
	SPA	20–25
	SPB	25–30
	SPC	40–50
Ribbed belt	J	8–20
	L	22–30
	M	38–80
Flat belt		2–3% L
Synchronous	P, mm = 3	10–16*
	5	16–21
	8	24–35
	14	38–60
	20	49–80

*For flanged pulleys. For unflanged pulleys, 2–5 mm is sufficient for all sections

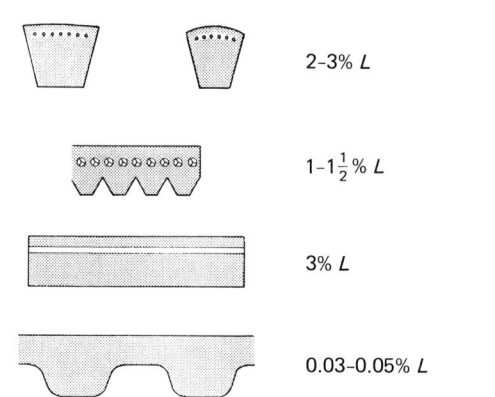

L = total belt length

Take-up allowance

Belt type	Allowance
V-belt / Wedge	2–3% L
Ribbed	1–1½% L
Flat	3% L
Synchronous	0.03–0.05% L

DESIGNING MULTI-DRIVE SYSTEMS

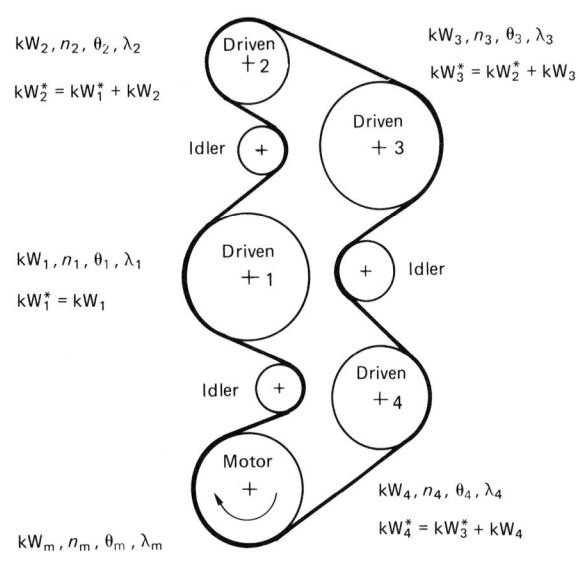

$kW_2, n_2, \theta_2, \lambda_2$
$kW_2^* = kW_1^* + kW_2$

$kW_3, n_3, \theta_3, \lambda_3$
$kW_3^* = kW_2^* + kW_3$

$kW_1, n_1, \theta_1, \lambda_1$
$kW_1^* = kW_1$

$kW_4, n_4, \theta_4, \lambda_4$
$kW_4^* = kW_3^* + kW_4$

$kW_m, n_m, \theta_m, \lambda_m$
$kW_m^* = kW_m$

For each driven shaft and the motor shaft, write down its transmitted design kW, its rev/min n, its approximate arc of contact θ, its recommended λ (from Figure 1.10) and kW*, the accumulated kW, as illustrated.

Select belt section from the power rating charts for the most severe combination at any shaft of kW* and n. For speed reduction drives, the motor shaft will always be the most severe.

Estimate the number of belts (or belt width) and belt tension as for a 2-shaft drive based on kW, n, d, λ for the motor pulley.

The resulting tension may not prevent slip on one or more of the driven shafts. Check this by calculating, for each driven shaft, i, and the motor shaft m, the quantities f_i and f_m below. Select the largest ratio of f_i/f_m. If this is greater than 1.0, increase the number of belts (or belt width) and belt tension by this factor.

It may be advantageous to drive the shafts by separate belts. The complexity may be such that a specialist should be consulted, particularly if any $\theta < 90°$ or if, for a synchronous belt, less than 6 teeth are in mesh on any pulley.

$$f_i = kW_i \frac{1+\lambda_i}{\lambda_i} - 2 kW_i^* \qquad f_m = kW_m \frac{1-\lambda_m}{\lambda_m}$$

Roller chain drives 2

SELECTION OF CHAIN DRIVES

The following selection procedure gives guidance for Industrial applications for the selection of chain drives comprised of roller chains and chain wheels conforming to ISO 606.

Chain selected using this method will have a life expectancy, with proper installation and lubrication, of 15 000 hours.

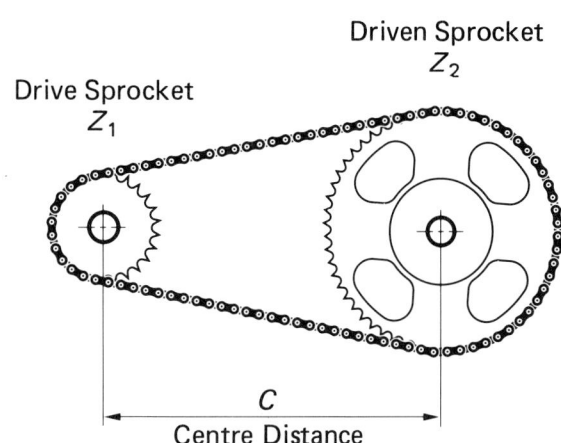

Z_1 – No of teeth on drive sprocket
Z_2 – No of teeth on driven sprocket
C – Centre Distance   mm
P – Chain Pitch   mm
i = Drive Ratio
L = Chain Length   Pitches

Basic information required

In order to select a chain drive the following essential information must be known:-
(a) The power, in kilowatts, to be transmitted.
(b) The speed of the driving and driven shafts in rev/min.
(c) The characteristics of the drive.
(d) Centre Distance.

From this basic information, the driver sprocket speed and selection power to be applied to the ratings charts, are derived.

Table 2.1 Drive ratios i, relative to one, using preferred sprockets

No. of teeth on driven sprocket Z_2	No. of teeth on drive sprocket Z_1					
	15	17	19	21	23	25
25	–	–	–	–	–	1,00
38	2,53	2,23	2,00	1,80	1,65	1,52
57	3,80	3,35	3,00	2,71	2,48	2,28
76	5,07	4,47	4,00	3,62	3,30	3,04
95	6,33	5,59	5,00	4,52	4,13	3,80
114	7,60	6,70	6,00	5,43	4,96	4,56

Step 1 Select drive ratio and sprockets

Table 2.1 may be used to choose a ratio based on the standard wheel sizes available.

It is best to use an odd number of teeth combined with an even number of chain pitches.

Ideally, chain wheels with a minimum of 17 teeth should be chosen. If the chain drive operates at high speed or is subjected to impulsive loads, the smaller wheels should have at least 25 teeth and should be hardened.

It is recommended that chain wheels should have a maximum of 114 teeth.

For large ratio drives, check that the angle of lap on Z is not less than 120°.

Drive ratios can otherwise be calculated using the formula

$$i = \frac{Z_2}{Z_1}$$

Step 3 Select drive

From the rating charts, Figure 2.1 and Figure 2.2, select the smallest pitch of simple chain to transmit the SELECTION POWER at the speed of the driving sprocket Z_1.

This normally results in the most economical drive selection.

Should the SELECTION POWER be greater than that shown, then consider a multiplex chain selected from the ratings charts.

Chain manufacturers should be consulted if any of the following apply.
(a) More than 1 driven sprocket
(b) Power or speeds above or below chart
(c) When volume production is envisaged
(d) Ambient conditions other than normal

Step 2 Calculate selection power

SELECTION POWER = Power to be transmitted $\times f_1 \times f_2$ (kW)

Where f_1 and f_2 are given in Tables 2.2 and 2.3

This selection power can then be used in Figure 2.1 and Figure 2.2 to select a suitable drive arrangement

This section has been compiled with the assistance of Renold Chains

2 Roller chain drives

Table 2.2 Application factor f_1

		Characteristics of driver		
		Smooth running	Slight shocks	Moderate shocks
	Driven machine characteristics	Electric motors. Steam and gas turbines. Internal combustion engines with hydraulic coupling.	Internal combustion engines with 6 cyls. or more with mechanical coupling. Electric motors with frequent starts. (2+ per day)	Internal combustion engines with less than 6 cyls. with mechanical coupling.
Smooth running	Centrifugal pumps and compressors. Printing machines. Paper calenders. Uniformly loaded conveyors. Escalators. Liquid agitators and mixers. Rotary driers. Fans.	1.0	1.1	1.3
Moderate shocks	Pumps and compressors (3+ Cyls). Concrete mixing machines. Non uniformly loaded conveyers. Solid agitators and mixers.	1.4	1.5	1.7
Heavy shocks	Planers. Excavators. Roll and ball mills. Rubber processing machines. Presses and shears. 1 & 2 cyl pumps and compressors. Oil drilling rigs.	1.8	1.9	2.1

Factor f_1 takes account of any dynamic overloads depending on the chain operating conditions. The value of factor f_1 can be chosen directly or by analogy using Table 2.2.

Table 2.3 Tooth factor f_2 for standard wheel sizes

Z_1	f_2
15	1.27
17	1.12
19	1.0
21	0.91
23	0.83
25	0.76

The tooth factor f_2 allows for the choice of a smaller diameter wheel which will reduce the maximum power capable of being transmitted, since the load in the chain will be higher.

Tooth factor f_2 is calculated using the formula

$$f_2 = \frac{19}{Z_1}$$

(Note that this formula arises due to the fact that selection rating curves shown in Figure 2.1 and Figure 2.2 are those for a 19 tooth wheel).

Recommended centre distances for drives

Pitch	(in)	$\frac{3}{8}$	$\frac{1}{2}$	$\frac{5}{8}$	$\frac{3}{4}$	1	$1\frac{1}{4}$	$1\frac{1}{2}$	$1\frac{3}{4}$	2	$2\frac{1}{2}$	3
	(mm)	9,525	12,70	15,875	19,05	25,40	31,75	38,10	44,45	50,80	63,50	76,20
Centre Distance	(mm)	450	600	750	900	1000	1200	1350	1500	1700	1800	2000

Chain length calculation

To find the chain length in pitches (L) for any contemplated centre distance of a two point adjustable drive use the following formula:-

$$\text{Length } (L) = \frac{Z_1 + Z_2}{2} + \frac{2C}{P} + \frac{\left(\frac{Z_2 - Z_1}{2\pi}\right)^2 \times P}{C}$$

The calculated number of pitches should be rounded up to a whole number of even pitches. Odd numbers of pitches should be avoided because this would involve the use of a cranked link which is not recommended.

If a jockey, or tensioner sprocket is used for adjustment purposes, two pitches should be added to the chain length (L).

C is the contemplated centre distance in mm and should generally be between 30–50 pitches. e.g. for a $1\frac{1}{2}$ P chain $C = 1.5 \times 25.4 \times 40 = 1524$ mm.

Selection of wheel materials

Choice of material and heat treatment will depend upon shape, diameter and mass of the wheel.

Pinion/wheel	Steady	Medium impulsive	Highly impulsive
Up to 29T	EN8 or EN9	EN8 or 9 hardened and tempered or case-hardened mild steel	EN8 or 9 hardened and tempered or case-hardened mild steel
30T and over	C.I.	Mild steel meehanite	Hardened and tempered steel or case-hardened mild steel or flame-hardened teeth

Through hardened EN9 should be oil quenched and tempered to give a hardness of 400 VPN nominal.

Roller chain drives 2

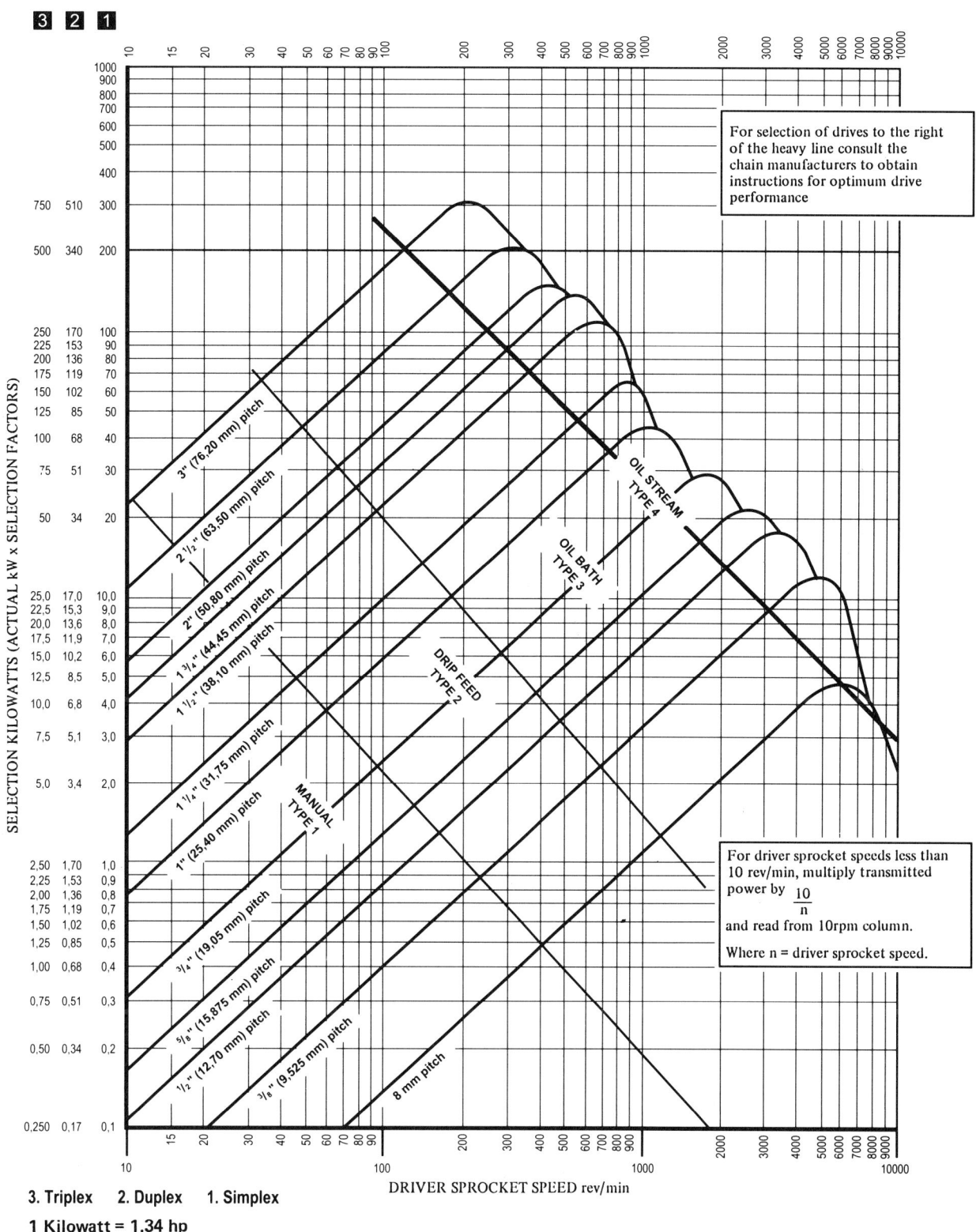

Figure 2.1 BS/DIN Chain drives – Ratings chart using 19T driver sprockets (ISO 606B)

2 Roller chain drives

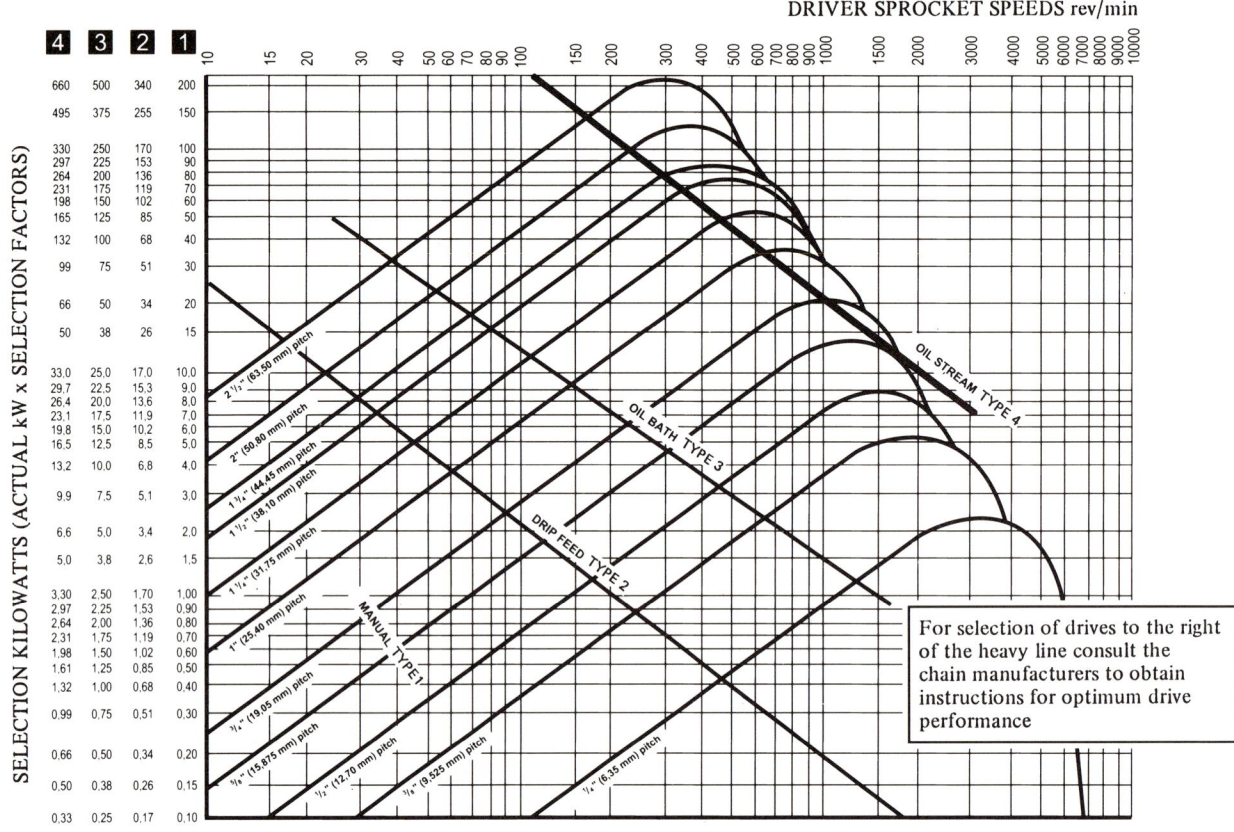

Figure 2.2 ANSI Chain drives – Ratings chart using 19T driver sprockets (ISO 606A)

LUBRICATION

Chain drives should be protected against dirt and moisture and be lubricated with good quality non-detergent petroleum based oil. A periodic change of oil is desirable. Heavy oils and greases are generally too stiff to enter the chain working surfaces and should not be used.

Care must be taken to ensure that the lubricant reaches the bearing area of the chain. This can be done by directing the oil into the clearances between the inner and outer link plates, preferably at the point where the chain enters the wheel on the bottom strand.

The table below indicates the correct lubricant viscosity for various ambient temperatures.

Recommended Lubricants.

Ambient temperature		Oil viscosity rating	
°C	°F (approx)	SAE	BS 4231
−5 to +5	20 to 40	20	46 to 68
5 to 40	40 to 100	30	100
40 to 50	100 to 120	40	150 to 220
50 to 60	120 to 140	50	320

For the majority of applications in the above temperature range a multigrade SAE 20/50 oil would be suitable.

Use of grease

As mentioned above, the use of grease is not recommended. However, if grease lubrication is essential the following points should be noted:
(a) Limit chain speed to 4 m/s.
(b) Applying normal greases to the outside of a chain only seals the bearing surfaces and will not work into them. This causes premature failure. Grease has to be heated until fluid and chain are immersed and allowed to soak until all air bubbles cease to rise. If this system is used the chains need regular cleaning and regreasing at intervals depending on power/speed.

Abnormal ambient temperatures

For elevated temperatures up to 250°C dry lubricants such as colloidal graphite or MoS_2 in white spirit or polyalkaline glycol carriers are most suitable.

Conversely, at low temperatures between −5 to −40, special low temperature initial greases and subsequent oil lubricants are necessary. Lubricant suppliers will give recommendations.

Roller chain drives

LUBRICATION METHODS

There are four basic methods for lubricating chain drives. The recommended methods shown in the ratings charts are determined by chain speed and power transmitted. The use of better methods is acceptable and may be beneficial.

Manual operation

Type 1
Oil is applied periodically with a brush or oil can, preferably once every 8 hours of operation. Volume and frequency should be sufficient to just keep the chain wet with oil and allow penetration of clean lubricant into the chain joints. Applying lubricant by aerosol can be satisfactory under some conditions, but it is important that the aerosol lubricant is of an approved type for the application. An ideal lubricant 'winds in' to the pin/bush/roller clearances, resisting both the tendency to drip or drain when the chain is stationary, and centrifugal 'flinging' when the chain is moving.

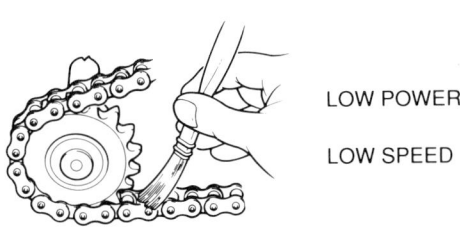

LOW POWER

LOW SPEED

Drip lubrication

Type 2
Oil drips are directed between the link plate edges from a drip lubricator. Volume and frequency should be sufficient to allow penetration of lubricant into the chain joints.

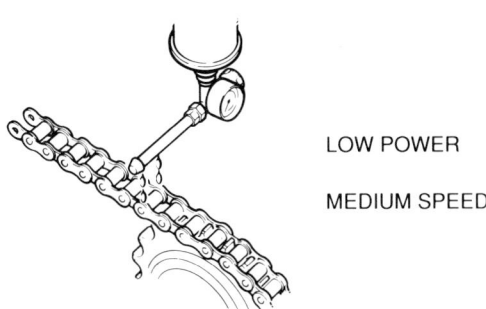

LOW POWER

MEDIUM SPEED

Bath or disc lubrication

Type 3
With oil bath lubrication the lower strand of chain runs through a sump of oil in the drive housing. The oil level should cover the chain at its lowest point whilst operating. With slinger disc lubrication an oil bath is used but the chain operates above the oil level. A disc picks up oil from the sump and deposits it on the chain by means of deflection plates. When such discs are employed they should be designed to have peripheral speeds between the minimum and maximum limits of 180 to 2440 m/min.

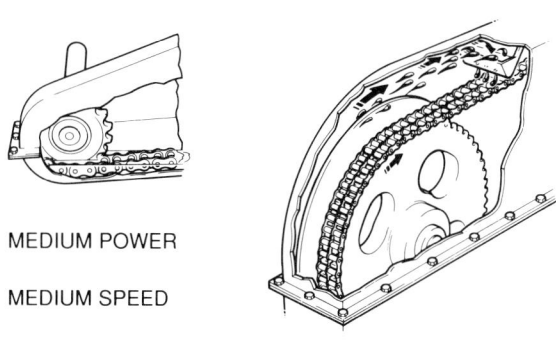

MEDIUM POWER

MEDIUM SPEED

Oil stream lubrication

Type 4
A continuous supply of oil from a circulating pump or central lubricating system is directed onto the chain. It is important to ensure that the spray pipe holes, from which the oil emerges, are in line with the chain plate edges. The spray pipe should be positioned so that the oil is delivered onto the chain just before it engages with the driver wheel. This ensures that the lubricant is centrifuged through the chain and assists in cushioning roller impact on the sprocket teeth. When a chain is properly lubricated a wedge of clean lubricant is formed in the chain joints and metal contact is minimised. Oil stream lubrication also provides effective cooling and impact damping at high speeds. It is, therefore, important that the method of lubrication specified in the ratings chart is closely followed.

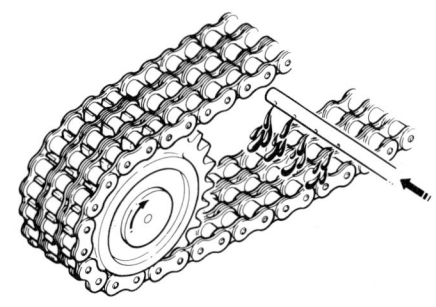

2   Roller chain drives

INSTALLATION AND MAINTENANCE

Wheel alignment

Make sure that the shafts are properly supported in bearings. Shaft, bearings and foundations should be suitable to maintain the initial static alignment.

Sprockets should be arranged close to the bearings. Accurate alignment of shafts and sprocket tooth faces provides uniform distribution of the load across the entire chain width and contributes substantially to maximum drive life.

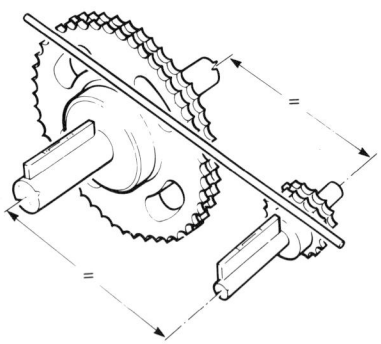

To measure wheel wear

Examination of the tooth flanks will give an indication of the amount of wear which has occurred. Under normal circumstances this will be evident as a polished worn strip about the pitch circle diameter of the sprocket tooth. If the depth of this wear has reached an amount equal to 10% of the 'Y'-dimension (see diagram) then steps should be taken to replace the sprocket. Running new chain on sprockets having this amount of tooth wear will cause rapid chain wear. It should be noted that in normal operating conditions with correct lubrication, the amount of wear at 'X' will not occur until several chains have been used.

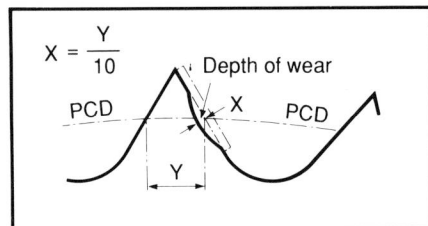

Health and safety

CAUTION
The following precautions must be taken before disconnecting and removing a chain from a drive prior to replacement, repair or length alteration:
1. Always isolate the power source from the drive or equipment.
2. Always wear safety glasses.
3. Always wear appropriate protective clothing, e.g. hats, gloves and safety shoes etc. as warranted by circumstances.
4. Always ensure tools are in good working condition and use in the proper manner.

Chain adjustment

The chain should be adjusted regularly so that, with one strand tight, the slack strand can be moved a distance 'A' at mid point (see diagram below). To cater for any eccentricities of mounting, the adjustment of the chain should be tried through a complete revolution of the large sprocket.

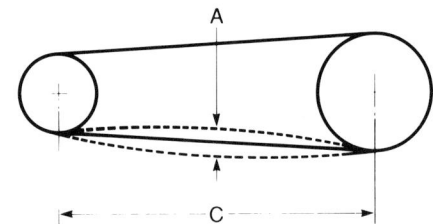

$A = \dfrac{C}{K}$ Where K = 25 for smooth drives
                     = 50 for impulsive drives

To measure chain wear

Measure length M in millimetres (see diagram). The percentage extension can then be calculated using the following formula:

$$\text{Percentage extension} = \frac{M - (X \times P)}{X \times P} \times 100$$

where X = number of pitches measured.
       P = Pitch in mm

As a general rule, the useful life of the chain is terminated, and the chain replaced, when the percentage extension reaches 2% (1% in the case of extended pitch chains). For drives with no provision for adjustment, the rejection limit is lower, dependent upon speed and layout. A useful figure is between 0.7% and 1% extension.

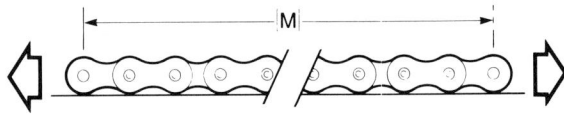

5. Always ensure that directions for the correct use of any tools are followed.
6. Always loosen tensioning devices.
7. Always support the chain to avoid sudden unexpected movement of chain or components.
8. Never attempt to disconnect or re-connect a chain unless the chain construction is fully understood.
9. Never re-use individual components.
10. Never re-use a damaged chain or chain part.
11. On light duty drives where a spring clip joint (No. 26) is used always ensure that the clip is fitted correctly in relation to direction of travel, with open end trailing.

Gears

GEAR TYPES

External spur gears

Cylindrical gears with straight teeth cut parallel to the axes, tooth load produces no axial thrust. Give excellent results at moderate peripheral speeds, tendency to be noisy at high speeds. Shafts rotate in opposite directions.

Internal spur gears

Provide compact drive for transmitting motion between parallel shafts rotating in same direction.

Helical gears

Serve same purpose as external spur gears in providing drive between two parallel shafts rotating in opposite directions. Superior in load carrying capacity and quietness in operation. Tooth load produces axial thrust.

Straight bevel gears

Used to connect two shafts on intersecting axes, shaft angle equals angle between the two axes containing the meshing gear teeth. Gear teeth are radial towards apex, end thrust is developed under tooth load tending to separate the gears.

Spiral bevel gears

Used to connect two shafts on intersecting axes same as straight bevels. Have curved oblique teeth contacting each other gradually and smoothly from one end of the tooth to the other. Meshes similar to straight bevel but are smoother and quieter in action. Have better load carrying capacity. Hand of spiral left-hand teeth incline away from axis in anti-clockwise direction looking on small end of pinion or face of gear, right hand teeth incline away from axis in clockwise direction. The hand of spiral of the pinion is always opposite to that of the gear and the hand of spiral of the pinion is used to identify the gear pair. The spiral angle does not affect the smoothness and quietness of operation or the efficiency but does affect the direction of the thrust loads created, a left hand spiral pinion driving clockwise when viewed from large end of pinion creates an axial thrust that tends to move the pinion out of mesh.

Zerol bevel gears

Zerol bevel gears have curved teeth lying in the same general direction as straight bevel gears but should be considered as spiral bevel gears with zero spiral angle.

3 Gears

Hypoid bevel gears

Hypoid gears are a cross between spiral bevel gears and worm gears, the axes of a pair of hypoid bevel gears are non-intersecting, the distance between the 'axes' being called the offset. The offset allows higher ratios of reduction than practicable with bevel gears. Hypoid gears have curved oblique teeth on which contact begins gradually and continues smoothly from one end of the tooth to the other.

Worm gears

Worm gears are used to transmit motion between shafts at right angles that do not lie in a common plane. They are also used occasionally to connect shafts at other angles. Worm gears have line tooth contact and are used for power transmission, but the higher the ratio the lower the efficiency.

APPLICATION OF GEARS

Table 3.1 Scope and torque capacity of gears

Relation between shaft axes	Gear ratio (RG)	Max. tooth speed V (M/Sec)	Type of tooth	Max. wheel torque (Nm)
Parallel	up to 10 to 1	5	Helical or straight	900×10^4
		25	Helical	220×10^4
			Profile ground straight	22×10^4
		205	Helical	56×10^4
Intersecting	up to 7 to 1	2.5	Spiral bevel or straight bevel	9×10^4
		60	Spiral bevel	4.5×10^4
Non-intersecting at 90°	up to 10 to 1	60	Hypoid bevel	6×10^4
Non-intersecting crossed at 90°	up to 50 to 1	50	Worm and wormwheel	28×10^4
			Crossed helical	17×10^4
Non-intersecting crossed at 80° to 100° but not 90°	up to 50 to 1	50	Worm and wormwheel	11×10^4
			Crossed helical	17×10^4

Note: The above figures are for general guidance only. Any case that approaches or exceeds the quoted limits needs special consideration of details of available gear-cutting equipment.

Gears 3

CHOICE OF MATERIALS

Table 3.2 Allowable stresses on materials for spur, helical, straight bevel, spiral bevel and hypoid bevel gears

Material	S_{CO} N/mm²	S_{BOS} (skin) N/mm²	S_{BOC} (core) N/mm²	BHN (core)	ULT (core) N/mm²	YIELD (core) N/mm²	VHN (skin) (min)	ULT (skin) N/mm²	BS spec.	Metallurgical condition and treatment
PB1	5	50	50	70	185	–	–	–	1400PB2	Sand cast
PB2	6	60	60	85	230	–	–	–	1400PB2	Chill cast centrifugal cast
PB3	7	70	70	90	260	–	–	–	1400PB2	
M1	6	76	76	140	310	–	–	–	309W24/8	Malleable iron
C1	7	40	40	165–190	185	–	–	–	1452	Ord. grade
C2	9	52	52	210–220	245	–	–	–	1452	Med. grade
C3	10	70	70	220–230	340	–	–	–	1452	High grade
C5	10	110	110	145–170	540	280	–	–	592	Steel as cast
1	10.5	145	145	150–180	540–618	280	–	–	BS 970 080A35	Normalised
2	12.5	160	160	180–230	618–740	430	–	–	150M28	Q
3	16.5	172.5	172.5	200–270	695–850	355	–	–	–	Normalised
4	18.5	186	186	220–270	740–895	585	–	–	708M40	S
5	18.5	186	186	220–270	740–895	585	–	–	817M40	S
6	21.5	215	215	250–300	850–1000	680	–	–	830M31	T
7	26.0	235	235	290–340	970–1130	770	–	–	830M31	V
8	45.0	110	145	150–180	540–618	–	450	1390	080A35	(FH)†
9	45.0	165	145	150–180	540–618	–	450	1390	080A35	(CH)†
10	45.0	130	172	200–250	695–850	–	450	1390	708M40	R(FH)†
11	45.0	195	172	200–250	695–850	–	450	1390	708M40	R(CH)†
12	45.0	145	186	220–270	740–895	–	450	1390	817M40	S(FH)†
13	45.0	207	186	220–270	740–895	–	450	1390	817M40	S(CH)†
14	45.0	240	215	250–300	850–1000	–	450	1390	826M31	T(CH)†
15	45.0	172	227	270–320	925–1005	–	450	1390	826M40	U(FH)†
16	45.0	255	227	270–320	925–1005	–	450	1390	826M40	U(CH)†
17	55.0	130	172	200–250	695–850	–	710	1850	722M24	R. Nitrided
18	62.0	145	186	223–270	740–895	–	710	1850	722M24	S. Nitrided
19	69.0	207	276	375–444	1205–1390	–	710	1850	897M39	Nitrided
20	69.0	255*	172	210–240	695–772	–	710	1850	665M17	Case-hardened
21	76.0	282*	214	250–300	740–1005	–	710	1850	655M13	
22	83.0	345*	262	350–410	1160–1312	–	710	1850	659M15	

Where:

S_{CO} = allowable contact stress
S_{BOS} = allowable skin bending stress
S_{BOC} = allowable core bending stress
CI: Cast iron
CS: Cast steel
BHN: Brinell hardness number
VHN: Vickers hardness number
MI: Malleable iron
PB: Phosphor bronze

*Multiply by 1.8 for very smooth fillets not ground after hardening.

(FH)†: Hardening by flame or induction over the whole working surfaces of the tooth flanks but excludes the fillets – applies to modules larger than 3.5.

(CH)† Hardening by flame or induction over the whole tooth flanks, fillets and connecting root surfaces – applies to modules between 5 and 28.
Spin hardening – applies to modules between 3.5 and 2.0.

Notes:

1. Materials 8 to 22, the basic allowable bending stress (S_{BO}), used in estimating load capacity of gears depends on the ratio of the depth of the hard skin at the root fillets to the normal pitch (circular pitch) of the teeth.

$$S_{BO} = S_{BOS} \text{ or } \frac{S_{BOC}}{[1 - 7.5 \text{ (depth of skin)/normal pitch}]}$$

whichever is the less.

2. Materials 8 to 22, values of S_{CO} are reliable only for skin thicker than:

$$0.003 \, d \times D \, (d + D)$$

3. Materials 1 to 8, the value of S_{BO} is approx:

$$S_{BO} = 600 \times \text{Ult Tensile} - \frac{\text{Ult. Tensile}^3}{60}$$

4. Gear cutting becomes difficult if BHN exceeds 270.

3 Gears

Table 3.3 Allowable stresses for various materials used for crossed helicals and wormwheels

S_{CO2} N/mm^2	S_{BO2} N/mm^2	Wheel material		BHN	Ultimate tensile strength N/mm^2
10.5	50	Phosphor	Sand cast	70	185–216
12.5	60	Bronze	Chill cast	82	230–260
15.2	70	BS 1400 P.B.2	Centrifugally cast	90	260–293
7	41.4	Cast iron	Ordinary grade	150	185–216
7	51.7	BS 1452	Medium grade	165	245–262
7	70.0		High grade	180	340–370

Note: The pinion or worm in a pair of worm gears should be of steel, materials 3 to 7 or 20 to 22, Table 3.2, and always harder than the material used for the wheel.

Non-metallic materials for gears

To help in securing quiet running of spur, helical and straight and spiral bevel gears fabric-reinforced resin materials can be used. The basic allowable stresses for these materials are approximately $S_{CO} = 10.5$ N/mm^2 and $S_{BO} = 31.0$ N/mm^2, but confirmation should always be obtained from the material supplier.

Other plastic materials are also available and information on their allowable stresses should be obtained from the material supplier.

Material combinations

1. With spur, helical, straight and spiral bevel gears, material combinations of cast iron – phosphor bronze, malleable iron – phosphor bronze, cast iron – malleable iron or cast iron – cast iron are permissible.
2. The material for the pinion should preferably be harder than the wheel material.

Where other materials are used:
(a) Where cast steel and materials 1 to 7, Table 3.2 are used, it is desirable that the ultimate tensile strength for the wheel should lie between the ultimate tensile strength and the yield stress of the pinion.
(b) Materials 8 to 22, Table 3.2, may be used in any combination.
(c) Gears made from materials 8 to 22, Table 3.2, to mate with gears made from any material outside this group, must have very smooth finish on teeth.

Gears 3

GEAR PERFORMANCE

A number of methods of estimating the expected performance of gears have been published as Standards. These use a large number of factors to allow for operational and geometric effects, and for new designs leave a lot to the designers' judgement, for the matching of the design to suit a particular application. They are, however, more readily applicable to the development of existing designs.

Early methods

Lewis Formula – Dates back to 1890s and is used to calculate the shear strength of the gear tooth and relate it to the yield strength of the material.

Buckingham Stress Formula – Dates back to mid 1920s and compares the dynamic load with the beam strength of the gear tooth, and a limit load for wear.

British Standard 436 Part 3 1986

Provides methods for determining the actual and permissible contact stresses and bending stresses in a pair of involute spur or helical gears.
   Factors covered in this standard include:

Tangential Force	The nominal force for contact and bending stress.
Zone factor	Accounts for the influence of tooth flank curvature at the pitch point on Hertzian stress.
Contact ratio factor:	Accounts for the load sharing influence of the transverse contact ratio and the overlap ratio on the specific loading.
Elasticity factor:	Takes into account the influence of the modulus of elasticity of the material and Poisson's ratio on the Hertzian stress.
Basic endurance limit:	The basic endurance limit for contact takes into account the surface hardness.
Material quality:	This covers the quality of the material used.
Lubricant influence, roughness and speed factor:	The lubricant viscosity, surface roughness and pitch line speed affect the lubricant film thickness which affects the Hertzian stresses.
Work hardening factor:	Accounts for the increase of surface durability due to meshing.
Size factor:	Covers the possible influences of size on the material quality and its response to manufacturing processes.
Life factor:	Accounts for the increase in permissible stress when the number of stress cycles is less than the endurance life.
Application factor:	This allows for load fluctuations from the mean load or loads in the load histogram caused by sources external to the gearing.
Dynamic factor:	Allows for load fluctuations arising from contact conditions at the gear mesh.
Load distribution:	Accounts for the increase in local load due to mal-distribution of load across the face of the gear caused by deflections, alignment tolerances and helix modifications.
Minimum demanded and actual safety factor:	The minimum demanded safety factor is agreed between the supplier and the purchaser. The actual safety factor is calculated.
Geometry factors:	Allow for the influence of the tooth form, the effect of the fillet and the helix angle on the nominal bending stress for application of load at the highest point of single pair tooth contact.
Sensitivity factor:	Allows for the sensitivity of the gear material to the presence of notches, ie: the root fillet.
Surface condition factor:	Accounts for the reduction of endurance limit due to flaws in the material and the surface roughness of the tooth root fillets.

3 Gears

International Standards Organisation I.S.O. 60 'Gears'

Similar in many ways to BS 436 Part 3 1986 but far more comprehensive in its approach. For the average gear design a very complex method of arriving at a conclusion similar to the less complex British Standard. Factors covered in this standard include:

Tangential load:	The nominal load on the gear set.
Application factor:	Accounts for dynamic overloads from sources external to the gearing.
Dynamic factor:	Allows for internally generated dynamic loads, due to vibrations of pinion and wheel against each other.
Load distribution:	Accounts for the effects of non-uniform distribution of load across the face width. Depends on mesh alignment error of the loaded gear pair and the mesh stiffness.
Transverse load distribution factor:	Takes into account the effect of the load distribution on gear-tooth contact stresses, scoring load and tooth root strength.
Gear tooth stiffness constants:	Defined as the load which is necessary to deform one or several meshing gear teeth having 1 mm face width by an amount of 1 μm.
Allowable contact stress:	Permissible Hertzian pressure on gear tooth face.
Minimum demanded and calculated safety factors:	Minimum demanded safety factor agreed between supplier and customer, calculated safety factor is the actual safety factor of the gear pair.
Zone factor:	Accounts for the influence on the Hertzian pressure of the tooth flank curvature at the pitch point.
Elasticity factor:	Accounts for the influence of the material properties, i.e.: modulus of elasticity and Poisson's ratio.
Contact ratio factor:	Accounts for the influence of the transverse contact ratio and the overlap ratio on the specific surface load of gears.
Helix angle factor:	Allows for the influence of the helix angle on surface durability.
Endurance limit:	Is the limit of repeated Hertzian stress that can be permanently endured by a given material.
Life factor:	Takes account of a higher permissible Hertzian stress if only limited durability endurance is demanded.
Lubrication film factor:	The film of lubricant between the tooth flanks influences surface load capacity. Factors include oil viscosity, pitch line velocity and roughness of tooth flanks.
Work hardening factor:	Accounts for the increase in surface durability due to meshing a steel wheel with a hardened pinion with smooth tooth surfaces.
Coefficient of friction:	The mean value of the local coefficient of friction is dependent on several properties of the oil, surface roughness, the 'lay' of surface irregularities, material properties of tooth flanks, tangential velocities, force and size.
Bulk temperature:	Surface temperature.
Thermal flash factor:	Dependent on moduli of elasticity and thermal contact coefficients of pinion and wheel materials, and the geometry of the line of action.
Welding factor:	For different tooth materials and heat treatments.
Geometrical factor:	Defined as a function of the gear ratio and a dimensionless parameter on the line of action.
Integral temperature criterion:	The integral temperature of the gears depends on the oil viscosity and the performance of the gear materials relative to scuffing and scoring.

The figures produced from this standard are very similar to those produced by British Standard 436 Part 3 1986.

Gears   3

British Standard 545, 1982 (Bevel gears)

Specifies tooth form, modules, accuracy requirements, methods of determining load capacity and material requirements for machine-cut bevel gears, connecting intersecting shafts which are perpendicular to each other and having teeth with a normal pressure angle of 20° at the pitch cone, whose lengthwise form may have straight or curved surfaces.

The load capacity of the gears is limited by consideration of both wear and strength, factors taken into account are:

Wear and strength factors:	Include the speed, surface stress, zone, pitch, spiral angle overlap ratio and bending stress factors.
Limiting working temperature:	The temperature of the oil bath under the specified loading conditions.
Basic stress factors:	Given for the various recommended materials.

British Standard 721 Part 2 1983 (Worm gears)

Specifies the requirements for worm gearing based on axial modules. Four classes of gear are specified, which are related to function and accuracy. The standard applies to worm gearing comprising cylindrical involute helicoid worms and wormwheels conjugate thereto. It does not apply to pairs of cylindrical gears connecting non-parallel axes known as crossed helical gears.

The load capacity of the gears is limited by both wear and strength of the wormwheel, factors taken into account include:

Expected life:	The strength is calculated to an expected total running life of 26 000 hours. Allows for both steady and variable loads at different running speeds.
Momentary overload capacity:	Momentary overload is considered as one whose duration is too short to be defined with certainty but does not exceed 15 seconds.
Efficiency and lubrication:	The efficiency, excluding bearing and oil-churning losses for both worm and wormwheel driving.
Basic stress factors:	Given for the various recommended materials.

American Gear Manufacturers Association Standards

The American Gear Manufacturers Association Standards are probably the most comprehensive coverage for gear design and are compiled by a committee and technical members representing companies throughout America, both north and south, Australia, Belgium, Finland, France, Great Britain, India, Italy, Japan, Mexico, Sweden, Switzerland and West Germany and are being constantly up-dated.

The standards cover gear design, materials, quality and tolerances, measuring methods and practices, and backlash recommendations.

Gear performance is covered by a series of different standards as follows:

AGMA 170–01	Design guide for vehicle spur and helical gears.
AGMA 210–02	Surface durability (pitting) of spur gear teeth.
AGMA 211–02	Surface durability (pitting) of helical and herringbone gear teeth.
AGMA 212–02	Surface durability (pitting) formulas for straight bevel and zerol bevel gear teeth.
AGMA 215–01	Information sheet for surface durability (pitting) of spur, helical, herringbone and bevel gear teeth.
AGMA 216–01	Surface durability (pitting) formulas for spiral bevel gear teeth.
AGMA 217–01	Information sheet – gear scoring design guide for aerospace spur and helical power gears.
AGMA 220–02	Rating the strength of spur gear teeth.
AGMA 221–02	Rating the strength of helical and herringbone gear teeth.

3 Gears

American Gear Manufacturers Association Standards (Contd.)

AGMA 222–02	Rating the strength of straight bevel and zerol bevel gear teeth.
AGMA 223–01	Rating the strength of spiral bevel gear teeth.
AGMA 225–01	Information sheet for strength of spur, helical, herringbone and bevel gear teeth.
AGMA 226–01	Information sheet – geometry factors for determining the strength of spur, helical, herringbone and bevel gear teeth.
AGMA 2001–B.88	Fundamental rating factors and calculation methods for involute spur and helical gear teeth.
AGMA 2005–B.88	Design manual for bevel gears.

Factors taken into account in these standards include:

Unit load:	Calculated from tangential load, size of gear teeth and face width of gear.
Bending stress factor:	The relation of calculated bending stress to allowable bending stress.
Geometry factor:	The geometry factor evaluates the radii of curvature of the contacting tooth profiles based on the tooth geometry.
Transmitted tangential load:	Represents the tooth load due to the driven apparatus.
Dynamic factors:	Account for internally generated tooth loads induced by non-conjugate meshing action of the gear teeth.
Application factor:	Allows for any externally applied loads in excess of the nominal tangential load.
Elastic coefficient:	Accounts for both the modulus of elasticity of the gears and Poisson's ratio.
Surface condition:	Allows for surface finish on the teeth, residual stress and the plasticity effects (work hardening) of the materials.
Size factor:	Reflects non-uniformity of material properties, tooth size, diameter of gears, ratio of tooth size to diameter, face width, area of stress pattern, ratio of case depth to tooth size and hardenability and heat treatment of materials.
Load distribution factors:	Modifies the rating equations to reflect the non-uniform distribution of the load along the lines of contact.
Allowable stress numbers:	Depend upon the material composition, mechanical properties, residual stress, hardness and type of heat treatment.
Hardness ratio factor:	Covers the gear ratio and the hardness of both pinion and gear teeth.
Life factor:	Adjusts the allowable stress numbers for the required number of cycles of operation.
Reliability factor:	Accounts for the effect of the normal statistical distribution of failures found in materials testing.
Temperatures factor:	Takes into account the temperature in which the gears operate.

Other factors are included in the standards depending upon the actual usage of the gears, e.g. motor vehicles, marine diesels, etc.

Comparison of design standards

From the list of factors given it can be seen that all three standards approach the gear performance problem in a similar manner but due to slight variances in methods used to calculate the factors the stress allowable figures will differ.

British Standard 436 Part 3 1986 is a radical up-date of the original BS 436 and in many ways brings it in line with ISO/TC 60 whilst it can be seen from the list of AGMA Standards that these are constantly reviewed to meet the demands of industry.

Flexible couplings 4

Flexible couplings are used to connect the shafts of separate machines and take up any small misalignment that there may be between them.

COUPLING TYPES

Gear couplings

Hubs fitted to the shafts of the coupled machines have gear teeth around their periphery and these mate with internal gear teeth on a sleeve which couples the two hubs together.

The gear meshes transmit the torque between the machines but allow relative movement to accommodate the misalignment.

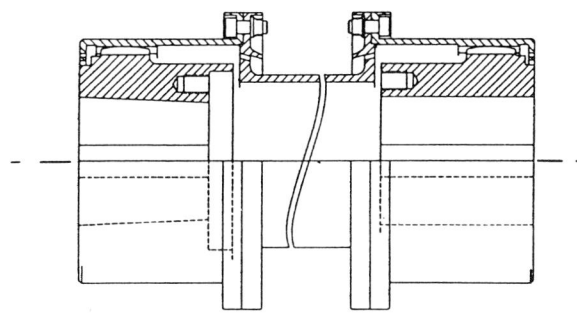

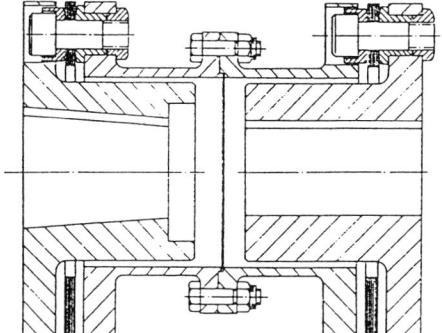

Multiple membrane couplings

The hubs on the shafts of the coupled machines are connected to an intermediate spacer by flexible members. These members are made from stacks of thin laminations so that they are flexible in bending and strong in tension and shear.

Some couplings use the flexible members as tangential links to provide tension connections. Other couplings use radial or disc shaped links in which the torque is transmitted in shear.

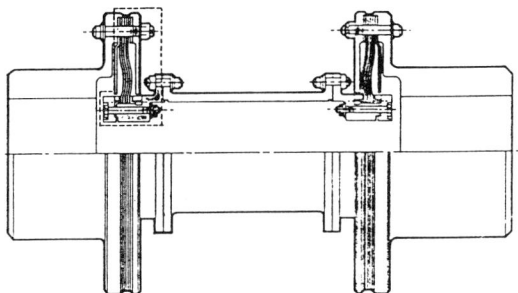

Contoured disc couplings

In this type of coupling the hubs on the machine shafts are coupled to an intermediate shaft by thin discs. These discs have a variable thickness in a radial direction to give a more even stress distribution and are made of high strength material.

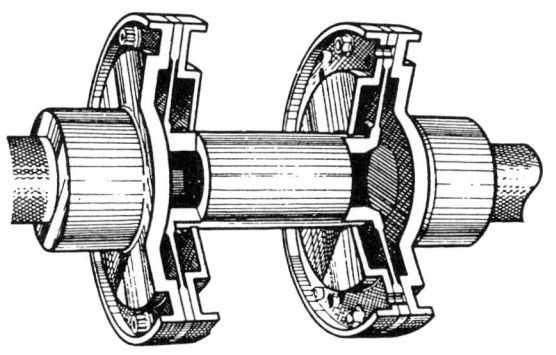

4 Flexible couplings

Elastomeric element couplings

There are various designs of coupling which use elastomeric materials to transmit the torque while allowing some flexibility.

The most highly rated couplings use the rubber mainly in compression in the form of rubber blocks located between radial blades. A hub with radial blades on its periphery is fitted to one machine, and a sleeve member with corresponding inwardly extending blades is fitted to the shaft of the other machine.

Convoluted axial spring couplings

The hubs on the coupled machines have a number of contoured blocks around their periphery. A convoluted axial spring is fitted around the hubs and into the slots between the blocks. The driving torque is transmitted by bending and shear in the axial bars of the spring, which can also deflect to take up misalignment.

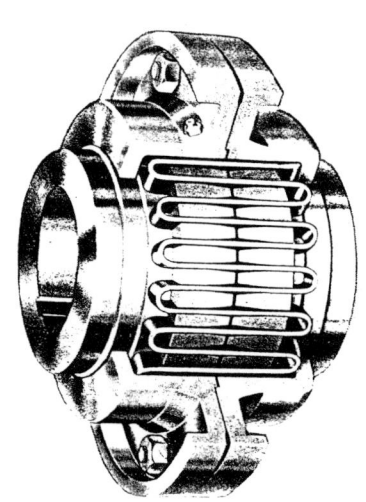

Quill shafts

The machines can also be coupled together by a shaft with a diameter which is just adequate to transmit the maximum torque, and made long enough to give lateral flexibility in order to take up misalignment. Quill shaft couplings do not permit any relative axial movement between the coupled machines.

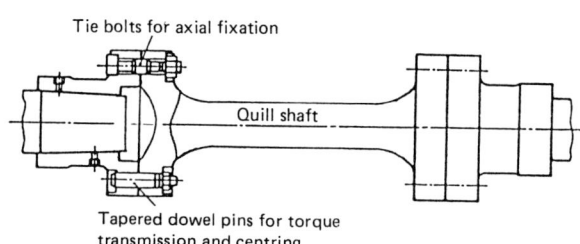

Flexible couplings   4

COUPLING PERFORMANCE

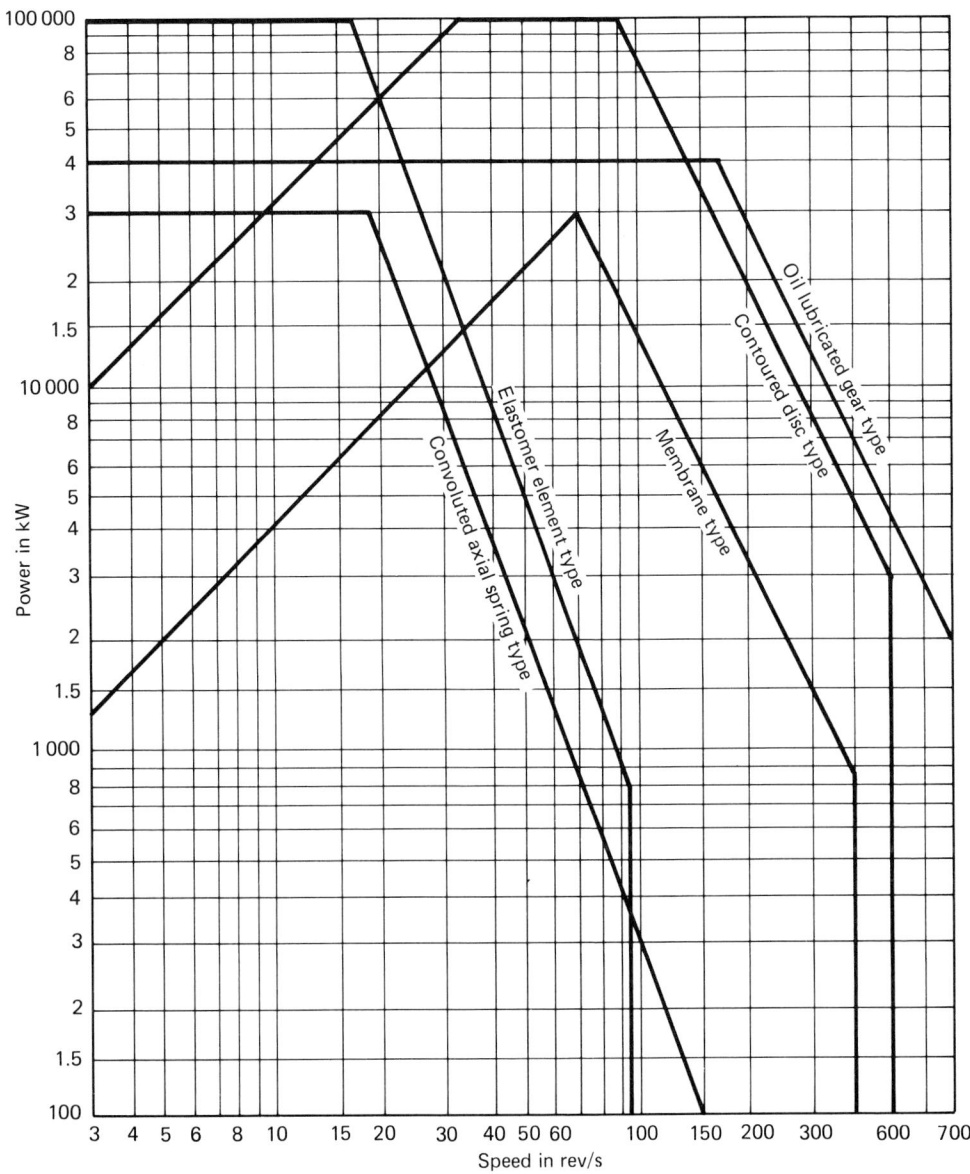

Figure 4.1 The power and speed limits of flexible couplings

In this figure the performance limits at higher speeds are determined largely by centrifugal stresses in the components.

The maximum power limits of gear couplings and elastomeric element couplings arise from contact and compressive stresses.

The lower speed performance limits of disc and membrane couplings arise from the maximum torque that can be transmitted within the stress limits of the material of the flexible members.

4 Flexible couplings

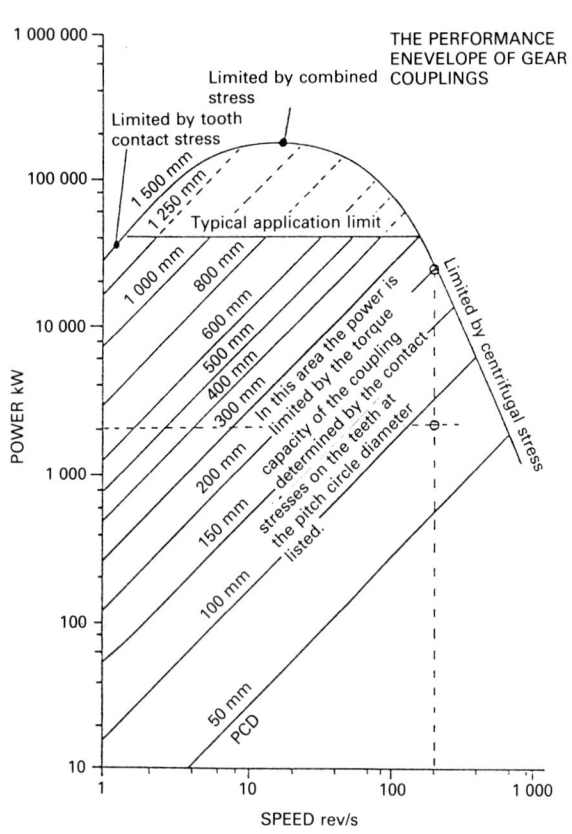

Figure 4.2 *The performance envelope for gear couplings*

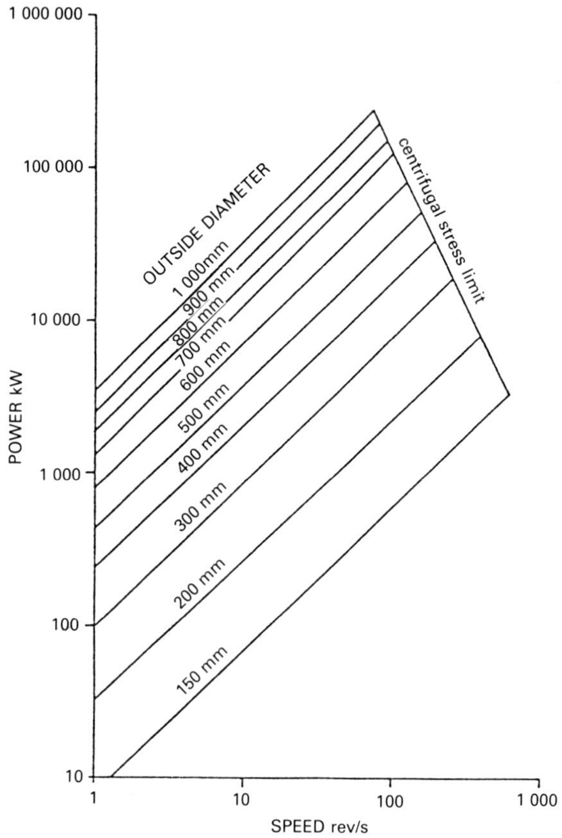

Figure 4.3 *Performance envelope for contoured disc couplings*

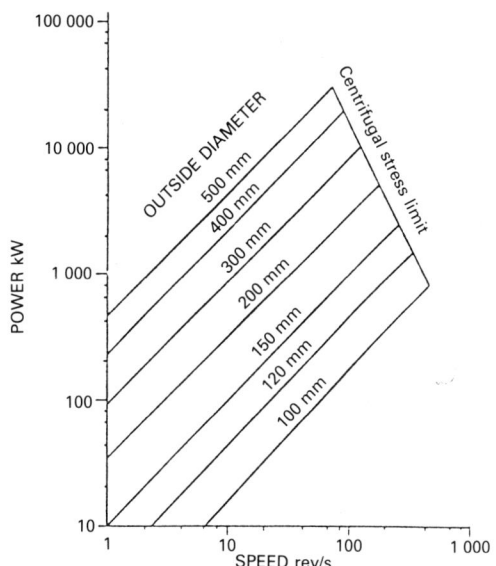

Figure 4.4 *Performance envelope for multiple membrane couplings*

The parallel inclined lines in these figures correspond to lives of constant torque. In these areas the performance of the coupling is limited by component stresses, which arise directly from the transmitted torque.

Flexible couplings 4

Table 4.1 Relative advantages and disadvantages of the various types of coupling

Coupling type	Advantage	Disadvantages
Gear couplings	Of the couplings that offer a range of flexibilities, they are the most compact and can provide the minimum overhung mass on the machine shaft. They allow the maximum axial movement of any coupling.	Lubrication is essential and while they can be packed with grease for low speed applications a continuous oil feed is required at high speeds. See Figure 4.5 for lubrication limits. They can apply substantial axial and lateral loads to the coupled machines. See Figure 4.7.
Multiple membrane couplings	They require no lubrication or maintenance and once correctly assembled should maintain their balance.	Their relatively high mass can affect the lateral stability of machine rotors. The diaphragm clamping is a critical assembly feature.
Contoured disc couplings	They require no lubrication and have inherently good balance. Their performance is predictable and consistent.	Have a large diameter which can give rise to windage losses and noise.
Elastomeric element couplings	Robust. Can absorb torsional shocks. Can be designed to de-tune torsional resonances in machine systems.	Relatively large diameter which limits their maximum speed capability.
Convoluted axial spring couplings	Robust with some torsional shock absorbing capability. Decoupling is simple, by removing covers and the convoluted spring.	Require grease lubrication which together with balance consistency, limits their maximum speed capability.
Quill shafts	Simplicity. Low mass. Balance retention	No axial movement capability and limited lateral flexibility.

Table 4.2 Approximate maximum misalignments allowable across the various types of coupling

Coupling type	Lateral misalignment	Axial misalignment
Gear couplings	Typically 0.002 radians per mesh but depends on diameter and rotational speeds, as it is limited by the rubbing speed at the teeth. See Figure 4.6.	Limited only by the axial width of the widest tooth row.
Multiple membrane couplings	Typically up to 0.008 radians/disc but depends on design and operating conditions. High axial displacements reduce the allowable angular misalignment.	Typically up to ±6 mm but depends on design.
Contoured disc couplings	Typically up to 0.010 radians/disc but depends on design and operating conditions.	Typically up to ±6 mm but depends on design.
Electromeric element couplings	Typically 0.008 radians and 1 mm laterally, but depends on design.	Typically up to 4 mm but depends on design.
Convoluted axial spring couplings.	Up to 0.2 mm laterally.	Up to 10 mm approximately.
Quill shafts	Depends on design possibly 0.002 radians along the length.	None unless used in conjunction with a disc coupling at one end.

4 Flexible couplings

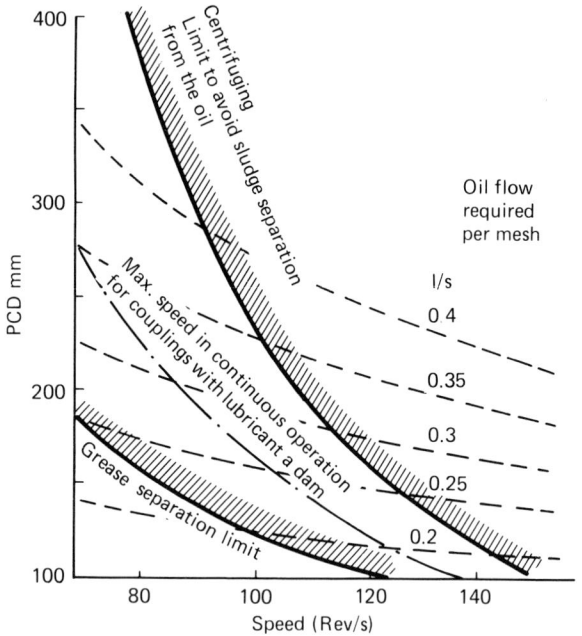

Figure 4.5 The lubrication requirements of gear couplings

Note: Damless coupling designs, with continuous lubricant feed to each tooth must be used for applications above the oil centrifuging limit and it is recommended that they should also be used above the chain dotted line where long periods of unattended reliable operation are required.

Flow rates indicated are for this type of design.

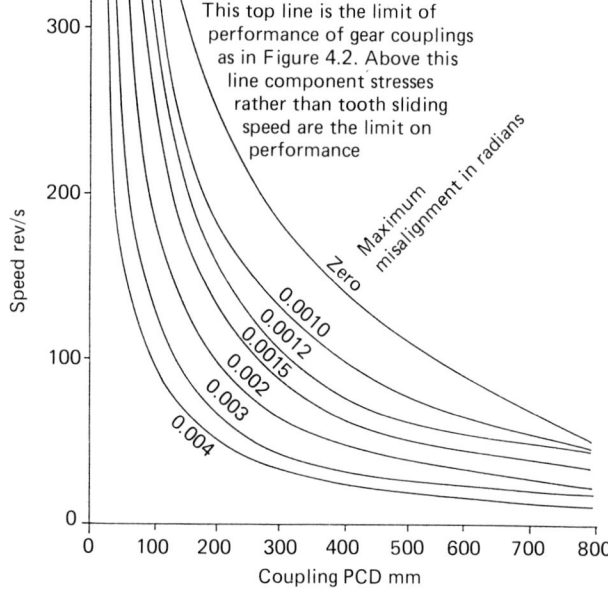

Figure 4.6 The maximum misalignment at a gear coupling mesh to avoid excessive tooth wear

Note
1. 0.001 radians of angular misalignment is equivalent to 0.001 inches per inch or 1 mm per metre misalignment across the mesh.
2. The maximum angular misalignments given are for continuous misaligned operation. If the misalignment is only transient, values up to 1.6 times greater are permissible.
3. The lines are plotted for a constant tooth sliding velocity of 0.12 m/s

Flexible couplings   4

COUPLING EFFECTS

Effect on critical speeds

Since couplings are fitted at the end of machine shafts, they constitute an overhung mass. Overhung masses reduce the lateral critical speed of rotors. If a machine is operating near its critical speed the overhung mass of the coupling needs to be considered.

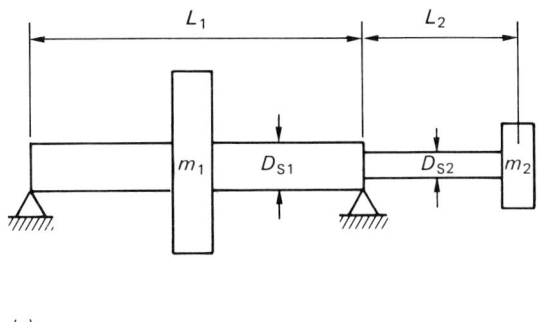

(a)

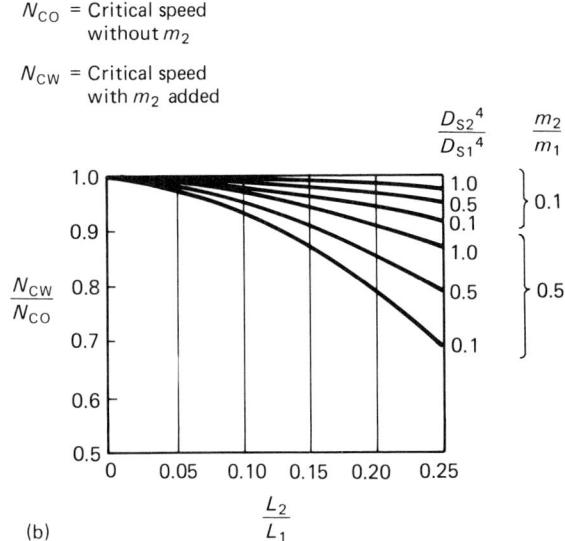

N_{CO} = Critical speed without m_2

N_{CW} = Critical speed with m_2 added

(b)

Gear coupling effects

The axial loads that may be applied by gear couplings to the coupled machines can be estimated from the tooth contact forces generated from the torque transmission multiplied by the likely coefficient of friction. This will normally have a value of about 0.15 but if the surface of the teeth becomes damaged could rise to 0.3.

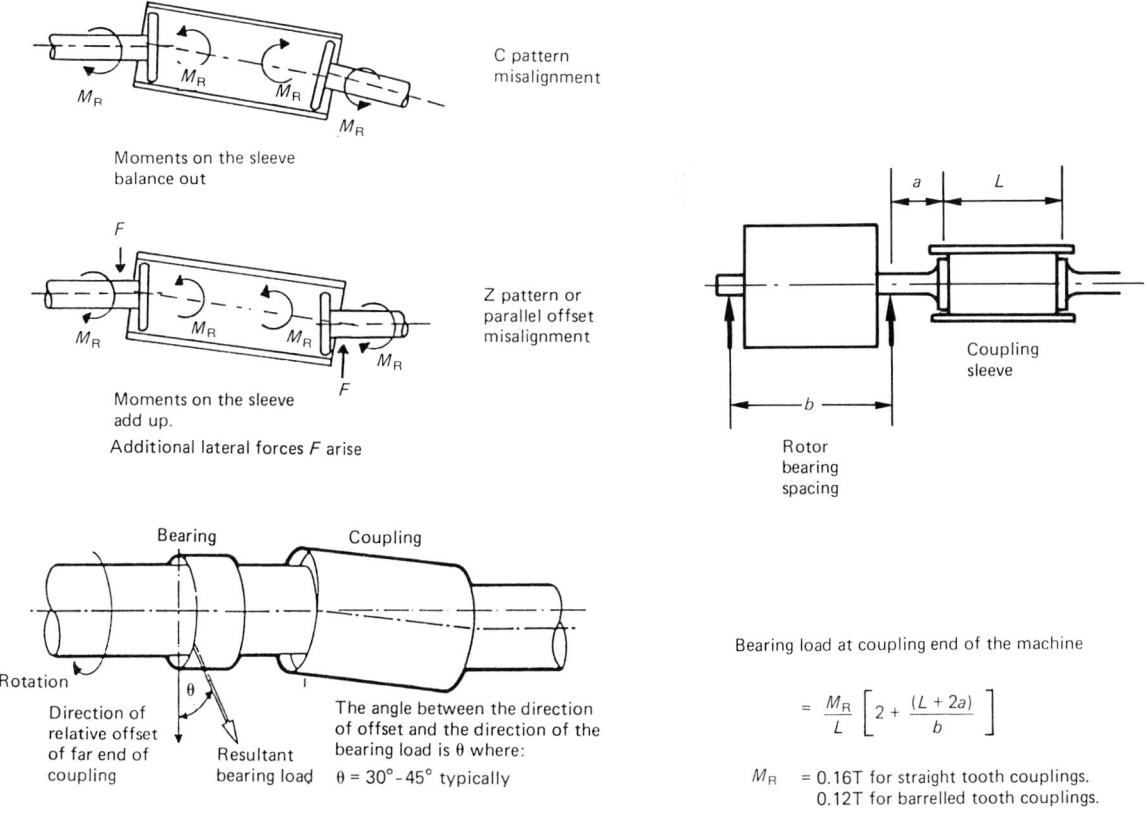

C pattern misalignment

Moments on the sleeve balance out

Z pattern or parallel offset misalignment

Moments on the sleeve add up.
Additional lateral forces F arise

The angle between the direction of offset and the direction of the bearing load is θ where:
θ = 30°–45° typically

Rotor bearing spacing

Bearing load at coupling end of the machine

$$= \frac{M_R}{L}\left[2 + \frac{(L + 2a)}{b}\right]$$

M_R = 0.16T for straight tooth couplings.
0.12T for barrelled tooth couplings.

Figure 4.7 Lateral bearing loads generated by gear couplings

5 Self-synchronising clutches

DESIGN AND OPERATION

Many machine systems require clutches which can pick up the drive to a machine that is already rotating or alternatively can release the drive when another driver takes over. For high power applications friction clutches tend to be large and generate heat due to slipping before synchronisation. Simple free wheels or sprag clutches are unsuitable for high torques because of their small driving contact areas.

High power overrunning clutches need to provide positive engagement and achieve a large driving surface area by transmitting torque through concentric internal and external teeth, generally gear teeth. When the clutch is disengaged these concentric teeth are separated axially.

A synchronising self-shifting clutch is shown in Figure 5.1. This has a pawl and ratchet mechanism to sense synchronism between the input and output shafts and to align the teeth which are then shifted into mesh by the small torque applied through the pawls to helical splines. Conversely reverse torque on the splines causes the teeth to disengage.

Powers of up to 300 MW have been transmitted by these clutches and the limiting factor in their design is centrifugal stresses in the outer geared rings.

High speeds in both the overrunning and engaged modes can be achieved (up to 15000 rev/min) because the driving teeth of the clutch are separated axially when the clutch is disengaged and because the pawl and ratchet mechanism can be designed for such high relative speeds without wear.

Clutches can be mounted directly between two shafts, in a separate casing or in a gearbox.

Oil lubrication is required for the clutch teeth and pawls to prevent wear and corrosion. Lower power and speed (e.g. 1 MW at 3000 rev/min) clutches are usually self-contained units with an integral oil system. Higher powers and speeds require a force fed oil supply, this is normally arranged to be common to the other machines in the system.

When the clutch is disengaged with the output side at high speed there will be a drag torque on the input shaft due to oil viscosity effects that may tend to keep the input machine rotating continuously at low speed. This can usually be accepted, if not a brake can be fitted to the clutch casing.

When the clutch teeth are engaged there is a strong centring effect between the input and output parts. If the input and output parts are always in good alignment (e.g. in a gearbox) this centring effect is acceptable, a semi-rigid clutch of this type is shown in Figure 5.2.

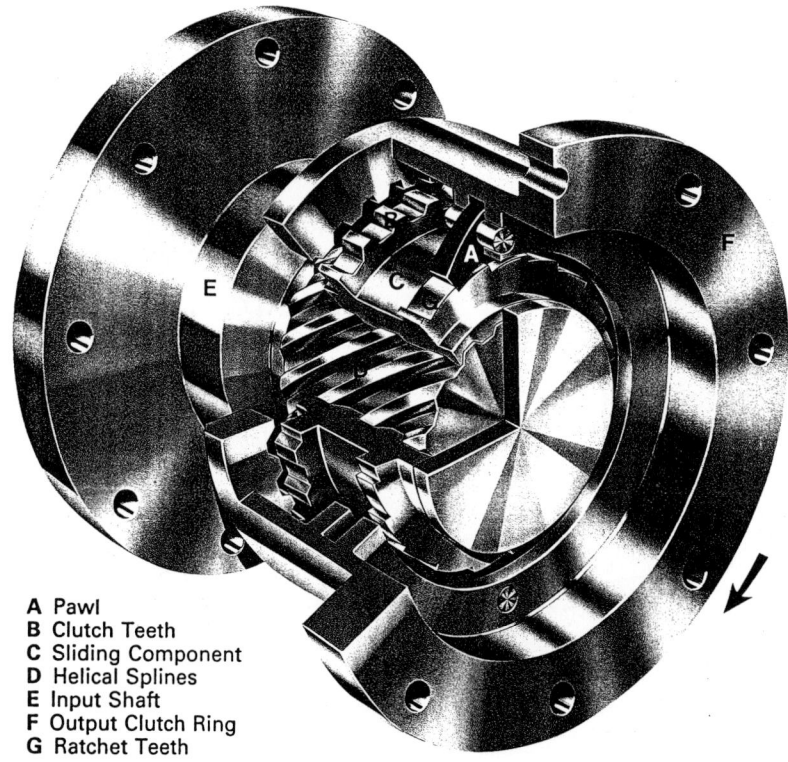

A Pawl
B Clutch Teeth
C Sliding Component
D Helical Splines
E Input Shaft
F Output Clutch Ring
G Ratchet Teeth

Figure 5.1 A high power clutch with axial engagement activated by helical splines and a ratchet and pawl mechanism

Self-synchronising clutches 5

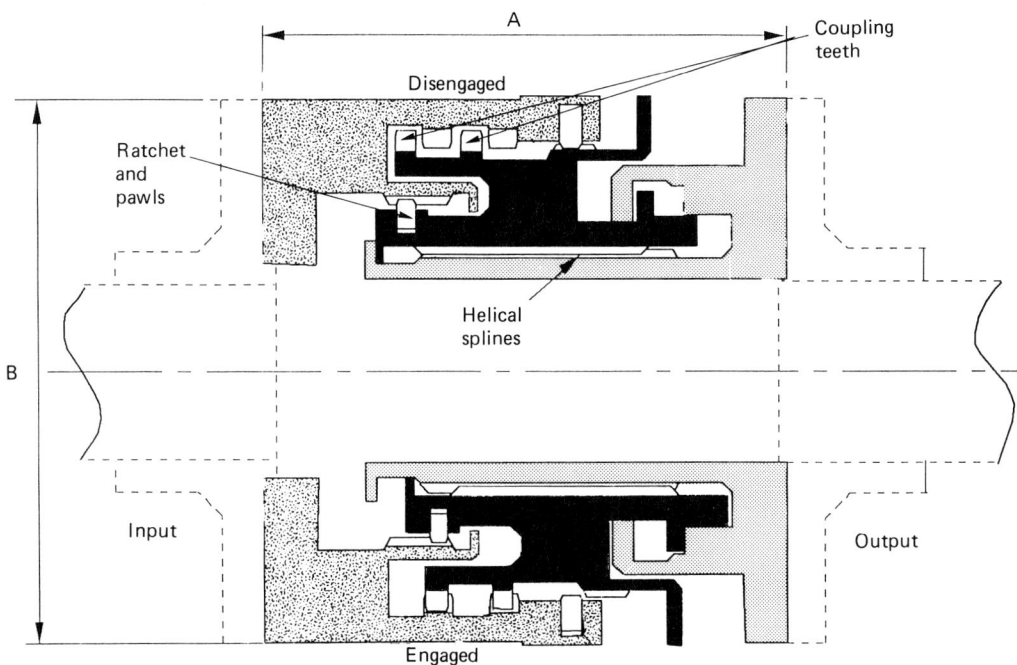

Figure 5.2 A high power semi-rigid clutch in the disengaged and engaged position

Clutches of this type, as shown in Figure 5.2, have typical dimensions and weights as shown in Table 5.1.

Table 5.1 Typical dimensions and weights of synchronising self shifting clutches

Torque capacity	A	B	Wt
40 kNm	360 mm	375 mm	150 kg
60	330	440	210
100	290	510	280
160	370	600	420
380	580	840	1125
1000	720	1000	2750

If there is misalignment between the input and output sides of the clutch (e.g: when a clutch is installed between separate machines such as a fan and a steam turbine) the clutch must be designed to accept this misalignment. Such

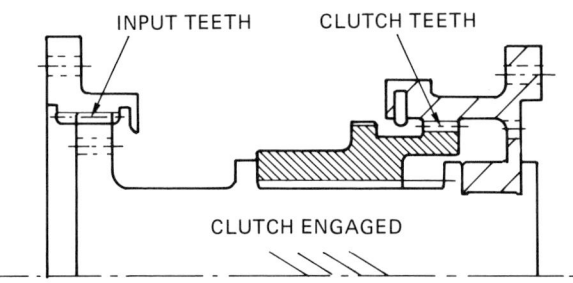

Figure 5.3 A spacer clutch which can also act as a flexible coupling between two shafts

a clutch acts like a spacer flexible coupling. A spacer clutch of this type is shown in Figure 5.3.

The clutch can also be designed to accept changes in axial length, as may be necessary due to the thermal expansion of a generator system.

In dual drive systems one driver can be stopped and maintenance can be carried out on that machine whilst the driven machine continues to run, driven by the other driver. To ensure the safety of operators the input side of the clutch must be held stationary by a maintenance lock which can be mounted within the clutch casing.

Clutches can be fitted with an internal thrust bearing between the input and output sides so that the clutch is a fixed length and positions the output machine from the input or vice versa. Such a bearing is usually only subjected to high thrust forces when the clutch is engaged and in this condition there is no relative rotation across the bearing so it has a high thrust capacity.

Clutches can be fitted with a lock so that once engaged and locked they do not disengage and hence can transmit reverse torque (e.g. in a marine reversing drive). The lock requires a control system to lock and unlock the clutch and this is normally arranged through a hydraulic servo interlinked with the plant control system. The lock can be preselected before clutch engagement so that the clutch locks immediately it engages.

The clutches can be fitted with a lock-out so that when selected the clutch will not engage (e.g. in a ship where an engine must be tested without driving the propeller). The lock-out again requires a control system.

Clutches can be arranged so that they only engage at high speeds for example in a steam turbine generator where the steam turbine must turn slowly forward for cooling without engaging the clutch and turning the generator.

5 Self-synchronising clutches

APPLICATIONS

Table 5.2 Applications and operating conditions of self synchronising clutches

Industrial drives		Powers	Speeds
Fan drive		500 kW	1500 rev/min
Pump drive		1000 kW	3000 rev/min
Compressor drive		3000 kW	6000 rev/min
Power generation			
Dual driven generator		2 × 50 MW	3000 rev/min
Combined cycle		100 MW/50 MW	3000 rev/min
Synchronous condensing		170 MW	3000 rev/min
Auxiliary drives			
Steam turbine barring		20 kW	10 rev/min
Gas turbine starting		200 kW	5000 rev/min
Marine drives			
Combined diesel or gas turbine propulsion (CODOG)		25 MW	3600 rev/min
		5 MW	1200 rev/min

Key to symbols

- Clutch
- F — Fan
- M — Motor
- GT — Gas turbine
- Geared drive
- C — Compressor
- P — Pump
- ET — Gas expander turbine
- HT — Hydraulic turbine
- G — Electric generator
- M/G — Motor/generator
- ST — Steam turbine
- Aero-derived gas turbine
- Diesel engine

One-way clutches 6

A freewheel or one-way clutch is a device which transmits torque in one direction, and disengages or freewheels in the other direction. The most familiar example is the bicycle freewheel which enables the cyclist to stop pedalling at will.

Numerous applications occur in small machinery. Two-speed drives can be designed with two motors of different speeds, coupled to the driven machine by one-way clutches. Indexing mechanisms using one-way clutches can convert reciprocating motion into unidirectional rotation. Uphill conveyors can be prevented from rolling back using a one-way clutch as a backstop.

Table 6.1 Types of one-way clutch

Type	Description	Special characteristics
Ratchet and pawl	Hard steel spring loaded pawls engage with ratchet teeth. (Springs not shown).	Backlash equal to ratchet tooth spacing. Makes clicking sound and suffers wear during freewheeling. Low precision, low cost. Plastic versions with axial configurations are possible.
Locking roller	Rollers ride up ramps and drive by wedging in place. Spring bias to engage. (Springs not shown).	Low cost because made by modifying roller bearings.
Locking needle roller	As for locking roller, but ramps are in the outer member. (Springs not shown).	Can operate directly against hardened shaft. Very compact unit, and low cost.
Sprag clutch	Profiled elements restrained by a cage. Jam in place to drive. Spring bias to engage. (Springs not shown).	Highest torque capability of all types. High cost. Can eliminate wear when freewheeling by means of centrifugal lifting of elements.
Wrap spring	Friction between hubs and coil spring causes spring to tighten onto hubs and drive in one direction. In opposite direction spring unwraps and slips easily.	Can be made very small. Potentially low cost but commercially available units are expensive.

6 One-way clutches

Table 6.2 Characteristics of one-way clutches

Type	Max torque	Max. freewheeling speed	Indexing accuracy	Cost
Ratchet and Pawl	10^3 Nm	low	poor	low
Roller clutch	5×10^4 Nm	moderate	good	low
Needle roller clutch	10^2 Nm	high	good	low
Sprag clutch	10^6 Nm	high	good	high
Wrap spring clutch	5×10^2 Nm	low	moderate	low/moderate

The above characteristics are for commercially readily available units. Larger, higher torque versions of all types are possible as special designs. For very high torques and power a better alternative is a self-synchronising clutch, described in section 5.

The graph below shows the approximate relationship between torque capacity and freewheeling speed for the various types.

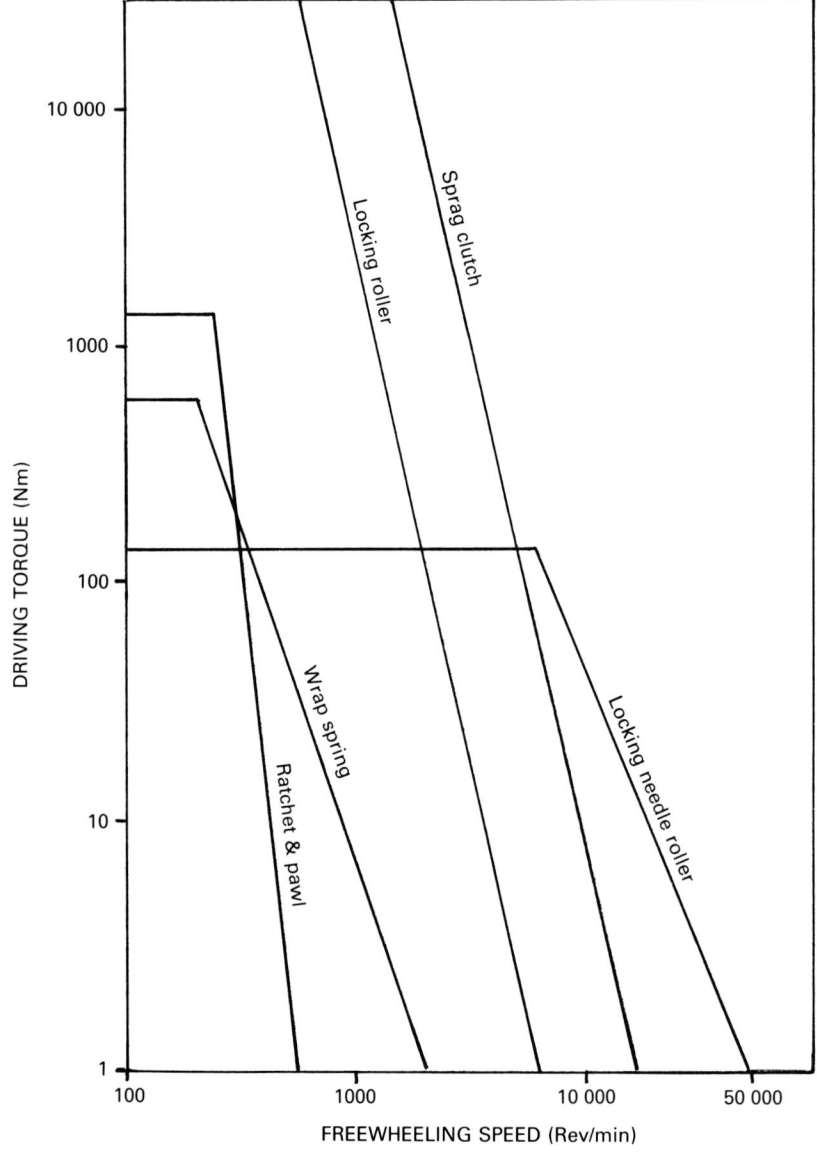

Figure 6.1 Approximate torque and freewheeling speed limitations for commercially available clutches

Friction clutches 7

CLUTCH SELECTION

A clutch is used to transmit motion from a power source to a driven component and bring the two to the same speed. Once full engagement has been made, the clutch must usually be capable of transmitting, without slip, the maximum torque that can be applied to it. The operating characteristics of the various clutch designs, and the requirements of the application, can be used as a guide to the selection of an appropriate type of clutch.

Table 7.1 Types of friction clutch

Type of clutch	Figure No.	Special characteristics	Typical applications
Cone type	7.1, 7.2	Embodies the mechanical advantage of the wedge which reduces the axial force required to transmit a given torque. It also has greater facilities for heat dissipation than a plate clutch of similar size and so may be more heavily rated	In general engineering its use is restricted to more rugged applications such as contractors' plant. Machine-tool applications include feed drives, and bar feed for auto. lathes. (Figure 7.2)
Single-plate (disc)	7.3	Used where the diameter is not restricted. Coil or diaphragm springs usually provide the clamping pressure by forcing the spinner plate against the driving plate. Simple construction, and if of the open type ensures no distortion of the spinner plate by overheating	Wide applications in automobile and other traction drives. Figure 7.5 (a and b) show alternative operating methods
Multi-plate	7.4	Main feature is that the power transmitted can be increased by using more plates, thus allowing a reduction in diameter. The spline friction should be minimised to ensure that clamping losses are small, and that the working rates of all plates are as uniform as possible. If working in oil,* it must be enclosed, whereas a dry plate clutch can often have circulating air to carry away the heat generated	Extensively used in machine tool head-stocks, or in any gearbox drive where space is limited between shaft centres. Figure 7.5(c) shows an operating method
Expanding ring or band	7.6, 7.7	Will transmit high torque at low speed. Centrifugal force augments gripping power, so withdrawal force must be adequate. Both cases show positive engagement	Large excavators. Textile machinery drives. Machine tools where clutch is located in main driving pulley
Centrifugal	7.8, 7.9	Automatic in operation, the torque without spring control increasing as the square of the speed. An electric motor with a low starting torque can commence engagement without shock, the clutch acting as a safety device against stalling and overload. Shoes are often spring-loaded to prevent engagement until 75% of full speed has been reached	Wide applications on all types of electric motor drives, generally reducing motor size and cost. Industrial diesel engine drives
Magnetic	7.10, 7.11, 7.12	Units are compact, and operate by a direct magnet pull through switch engagement. No end thrust is transmitted, and centrifugal force has no effect on drive. Stock clutches are generally wound for 24 V d.c. with powers up to about 37 kW at 100 rev/min. The response curves, (Figure 7.12) show typical torque and voltage time charts	Drives with automatic speed-changing systems requiring remote control. Machine tool gearboxes. N.C. machine tools. Tracer control systems for copying
Hydraulic	7.13	Clutches can be incorporated into automatic machine cycles by remote control of solenoid valves. They give high torque with minimum size, i.e. compared with magnetic clutches identical dimensions result in a torque ratio of 3:1. Maintenance low because of working in oil, and piston stroke compensates for wear	Marine reversing drives. Excavators. Diesel turbine drives. Winch drives. Presses
Pneumatic		Function similar to that of hydraulic clutch, producing axial pressure by cylinder and piston. General design is usually of the multi-disc type	Crank presses. Flying shear. Rolling mill drives. Cranes. Hammers

* Working in oil gives a reduction in friction, but this can be counteracted by higher operating pressures. As long as there is an oil film on the plates, the friction and the engagement torque remain low, but as soon as the film breaks the engagement torque rises rapidly and may lead to rapid acceleration. The friction surface pressure should not exceed 1 MN/m^2 with a sliding speed maximum of 20 m/s for steel on steel. With oil immersed clutches having steel and sintered plates, the relationship between static and dynamic coefficient of friction is more favourable. Friction surface pressure and sliding speed may then be up to 3 MN/m^2 and 30 m/s.

7 Friction clutches

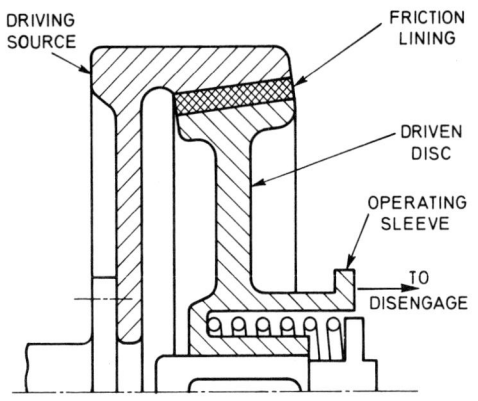

Figure 7.1 Cone-type friction clutch

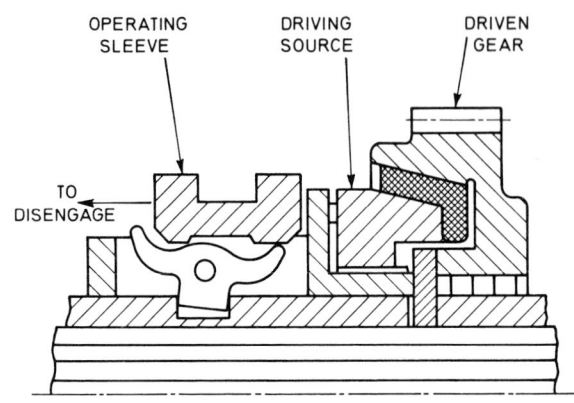

Figure 7.2 Cone clutch used on automatic lathe

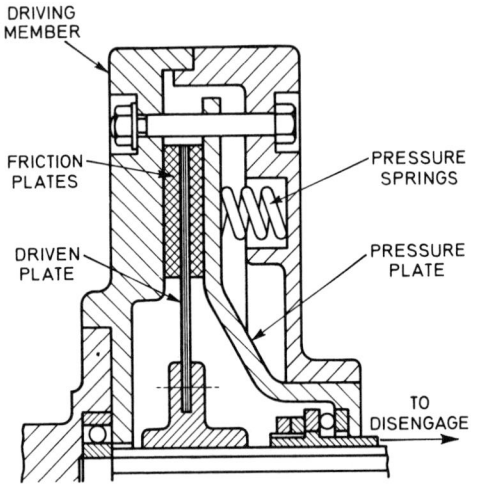

Figure 7.3 Construction of single-plate clutch

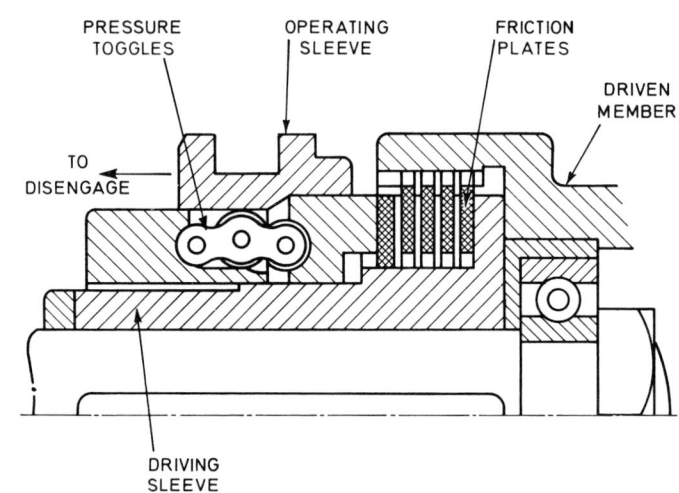

Figure 7.4 Design of multi-plate clutch

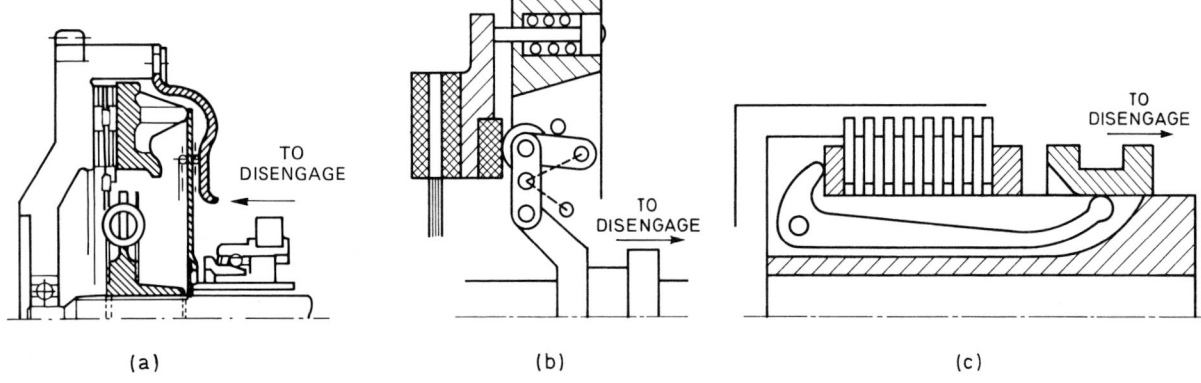

Figure 7.5 Alternative operating methods: (a) diaphragm spring clutch; (b) over-centre clutch, suitable for long engagement periods—no external force is required once clutch is engaged; (c) pivoted lever design for engagement of multi-plate clutch

Friction clutches 7

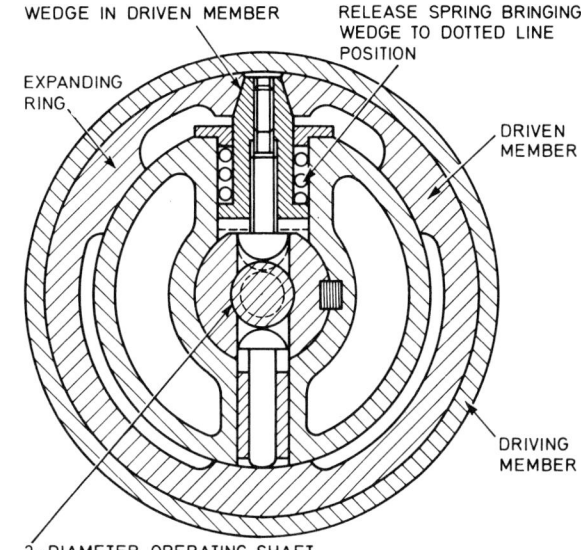

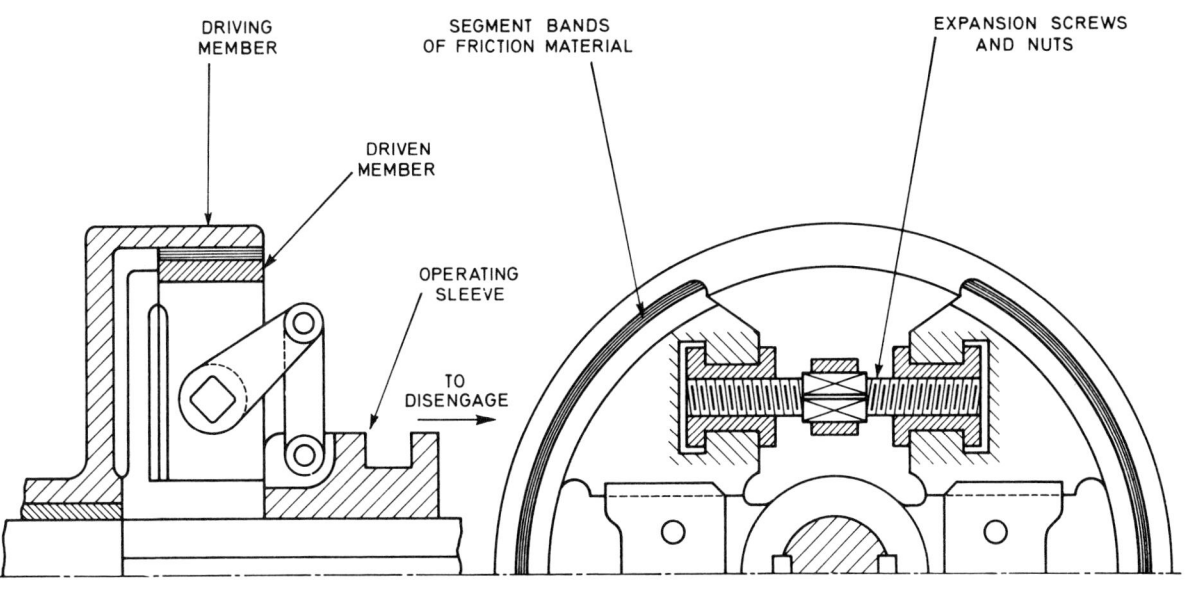

Figure 7.6 Expanding ring type of friction clutch

Figure 7.7 Expanding band friction clutch

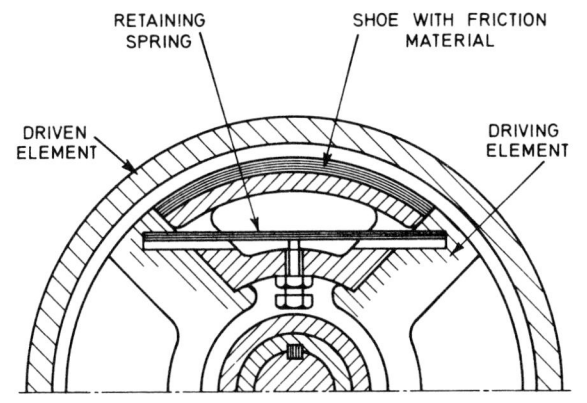

Figure 7.8 Centrifugal clutch with spring control

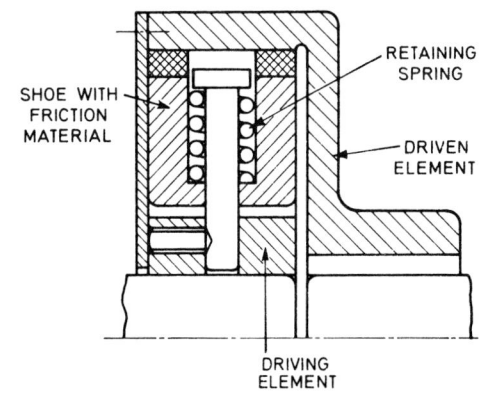

Figure 7.9 Light-duty type of centrifugal clutch

7 Friction clutches

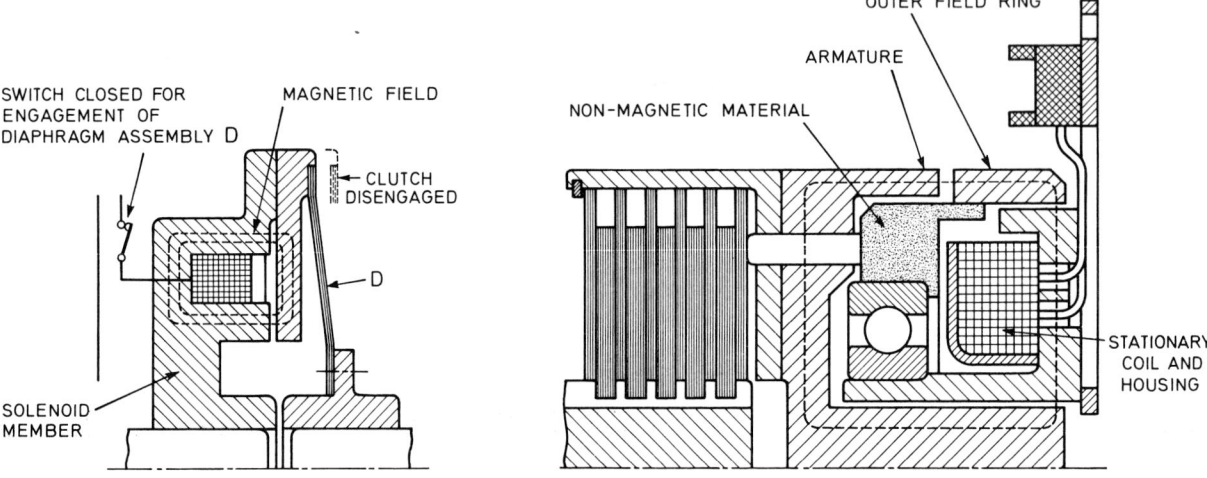

Figure 7.10 Diaphragm type of magnetic clutch

Figure 7.11 Multi-plate type of magnetic clutch

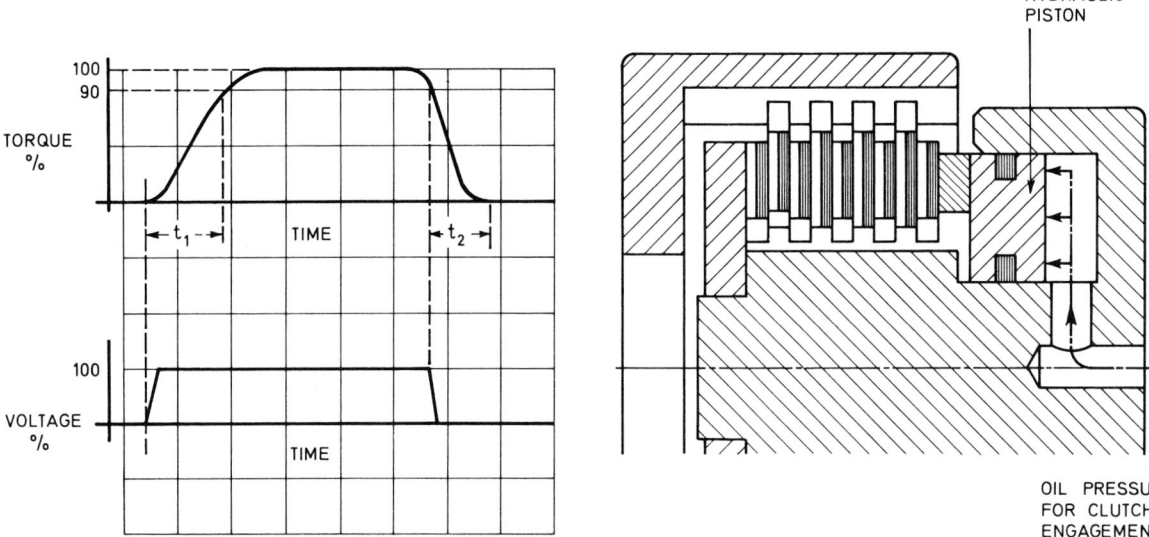

Figure 7.12 Response curves of magnetic clutch

Figure 7.13 Diagram showing operation of hydraulic clutch

Friction clutches 7

DESIGN OF CLUTCHES

Choice of lining area

The total lining area required, A, may be determined from: $A = H/K_l$, where H is the power to be dissipated, and K_l is the duty rating of the lining material.

The power to be dissipated, H, may be determined from the energy to be imparted to the driven component in the time of engagement or, if the transmitted torque is considered to be constant, from the transmitted torque and the mean relative speed during engagement.

When the total area is known, the detail dimensions of the linings can be readily calculated for plate or drum clutches. For cone clutches, the developed shape of the lining is shown in Figure 7.14 and can be calculated from

$$R = \frac{D}{2 \sin \theta}$$

$$\beta = 360° \sin \theta$$

$$A = 2R \sin \frac{\beta}{2}$$

$$B = R - (R - b) \cos \frac{\beta}{2}$$

The required service life and the wear rate and allowable wear of the lining are also important, particularly in determining the lining thickness.

Typical lining arrangements, contact pressures and suggested duty ratings for lining materials are given in Table 7.2.

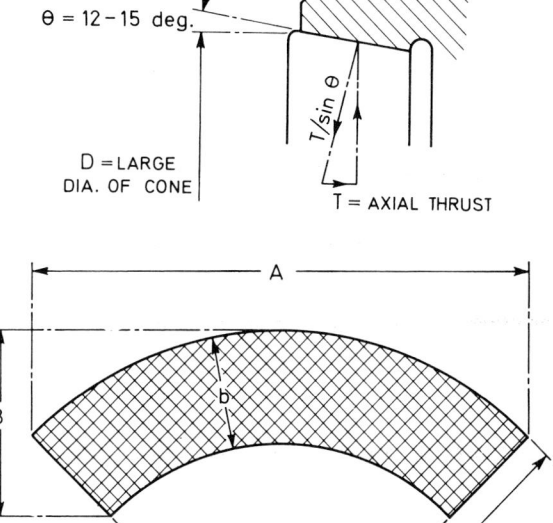

Figure 7.14 Development of liner for cone clutch

Table 7.2 Guide to design and material selection for various types of clutch

Type of friction device	Self-energisation and sensitivity	Angle of lining arc	Pressure range MN/m²	Material	Severity of the application	Suitable duty rating of lining, K_l W/mm²	Coefficient of friction
Two leading shoe drum clutch	High	100° per shoe	0.14–1.0	Moulded or woven	High	0.24	0.30
					Medium	0.58	0.35
					Low	1.75	0.40
Band clutch operating with drum rotation	Very high	270°	0.07–0.7	Rigid moulded segments, flexible moulded or woven	High	0.24	0.25
					Medium	0.58	0.30
					Low	1.15	0.35
Plate clutch operating under dry conditions	None	360°	0.07–0.4	Rigid woven or moulded	High	0.12	0.30
					Medium	0.24	0.35
					Low	0.58	0.40
Plate clutch operating under oil immersed conditions	None	360°	0.28–2.8	Rigid woven or moulded	High	0.15	0.06
					Medium	0.29	0.08
					Low	0.70	0.10
Cone clutch	None	360°	0.05–0.2	Moulded or woven	High	0.24	0.25
					Medium	0.35	0.30
					Low	0.80	0.35

7 Friction clutches

Lining design for oil-immersed clutches

For working in oil, multi-plate clutches are suitable. Oil acts as a cushion and energy released by heat is carried away by oil. The main disadvantage is a reduction in friction, but this can be counteracted by higher operating pressures. As long as there is an oil film on the plates, the friction characteristic and engagement torque remain low, but as soon as the film breaks the engagement torque rises rapidly and may lead to rapid acceleration. The friction surface pressure should usually not exceed 1 MN/m^2 with a sliding speed maximum of 20 m/s, steel on steel. With oil-immersed clutches having steel and sintered plates the relationship between the static and dynamic coefficient of friction is more favourable. Friction surface pressure and sliding speed may be up to 3 MN/m^2 and 30 m/s.

Facing grooves and anti-distortion slots

The most common facing groove is a single- or multi-lead spiral. This helps to prevent the formation of an oil film, which, if formed, would lower the coefficient of friction. It also provides space for the oil to be dispersed during clutch engagement. Spiral grooves are between 0.6 and 1.5 mm wide and 0.2–1 mm deep, depending on diameter and face thickness. The pitch is between 1.5 and 6.0 mm, depending on the size of the disc.

Figure 7.15 shows an anti-distortion slot in a sintered metal facing disc. Two widths of slots are frequently used, 4 or 5 mm wide. The slots minimise distortion and warping. Thermal expansion of the sintered metal and backing plates from absorption of heat causes expansion and contraction with temperature variations. The slots permit dimensional changes without buckling or dishing the backing plate. External and internal slots are used as in Figure 7.16. They number from 4 to 8 of each.

On moulded and woven fabrics the slots are normally 2.5–3.5 mm wide and 1–1.5 mm deep. Eighteen to twenty-four equi-distant slots or sets of three are usual. Thermal expansion slots are not required for these materials.

Clutch plates (Figure 7.17) are also available, which present a sine wave cross-section along a periphery, i.e. a corrugation, which generates a spring action. During engagement the pressure on the friction surface increases until the sine wave becomes a flat surface. The spring action ensures positive declutching and only line contact results in the disengaged position, thus ensuring minimum torque and heating when idling.

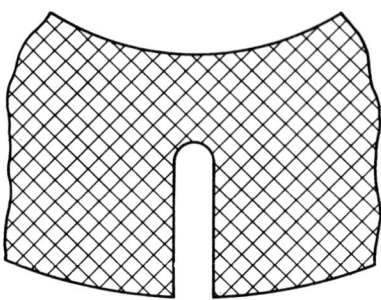

Figure 7.15 Anti-distortion slot in friction plate

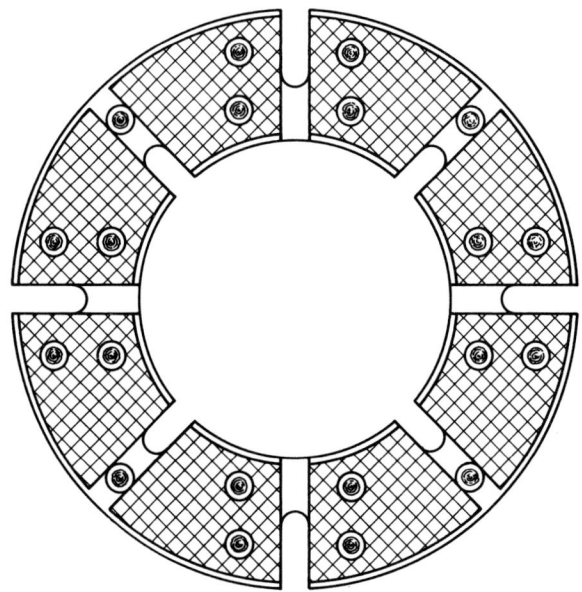

Figure 7.16 Slots to counteract expansion and contraction of plates

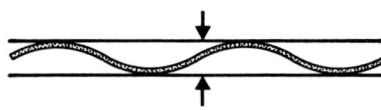

Figure 7.17 Friction plates with sine wave cross-section

Friction clutches 7

FITTING OF FRICTION LININGS

Copper or brass semi-tubular rivets are used for the attachment of the majority of the linings. One manufacturer uses brass containing 70% copper in 150° head semi-tubular rivets, as shown in Figure 7.18. The recommended dimensions and lining area/rivet are as follows:

Lining thickness (mm)	4.8	6.35	9.5	12.7	19.0
Rivet shank dia. (mm)	4.0	4.8	6.35	8.0	9.5
Lining area/rivet (mm^2)	1900	2300	3600	4500	6500

With riveting, some lining area is lost to rivet holes, and up to a third of the thickness is used to accommodate rivet heads, thus reducing wear life. Friction clutch facings, particularly those used on cone and band clutches, can be bonded to the metal carrier using proprietary adhesives and techniques (contact the manufacturers). Bonded facings have the advantage that all the friction material can be worn away.

Some precautions to observe when lining cone clutches are shown in Figure 7.19.

LINING MATERIALS

Impregnated woven cotton based linings are used to obtain high friction, but the maximum operating temperature is limited to that at which cotton begins to char (100°C), therefore asbestos and non-asbestos fibres have replaced cotton for applications where greater heat resistance is required. The fibres are woven to produce a fabric which is impregnated with a resin solution and cured. Zinc or copper wire is often introduced to increase thermal conductivity. Asbestos and non-asbestos moulded friction materials consist basically of a cured mix of short asbestos or other fibres and bonding resins and may also contain metal particles.

Asbestos or non-asbestos tape or yarn can be wound into discs and bound together using resin or rubber compounds.

Sintered metals are used for a limited number of friction applications. The metal base is usually bronze, to which is added lead, graphite and iron in powder form. The material is suitable for applications where very high temperatures and pressures are encountered. It is rigid and has a high heat conductivity, but gives low and variable friction.

Information on the various lining materials is given in Tables 7.3, 7.4 and 7.5.

Mating surfaces

The requirements are: (1) requisite strength and low thermal expansion; (2) hardness sufficient to give long wear life and resist abrasion; (3) heat soak capacity sufficient to prevent heat spotting and crazing.

Close-grained pearlitic grey cast iron meets these requirements, a suitable specification being an iron with the following percentage additions: 3.3 carbon, 2.1 silicon, 1.0 manganese, 0.3 chromium, 0.1 sulphur, 0.2 phosphorous, 4.0 molybdenum, 0.5 copper plus nickel. Hardness should ideally be in the range 200–230 BHN.

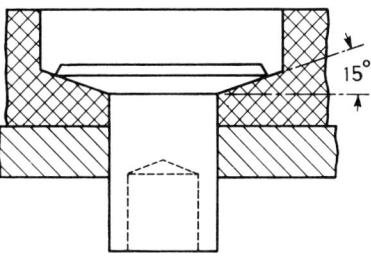

Figure 7.18 Type of rivet and fastening for clutch linings

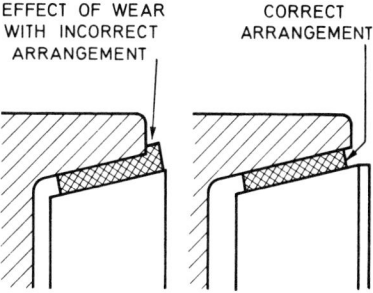

Figure 7.19 Precautions to take when lining cone clutches

7 Friction clutches

Table 7.3 Clutch facing materials and their applications

Type	Uses
Woven	Industrial band, plate and cone clutches—cranes, lifts, excavators, winches, machine tools and general engineering applications
Asbestos/non asbestos wound tape and yarn	Mainly automotive and light commercial vehicles, agricultural and industrial tractors
Moulded	Automotive, commercial vehicles, agricultural and industrial tractors
Sintered	Tractors, heavy vehicles, road rollers, winches, machine tool applications
Cermet	Heavy earth-moving equipment, crawler tractors, sweepers, trenchers and graders
Oil immersed	
paper	Automotive and agricultural automatic transmissions
woven	Band linings and segments for automatic transmissions
moulded	Industrial transmissions and agricultural equipment
sintered	Power shift transmission, presses, heavy-duty general engineering applications

Facings are available in a wide range of sizes 75–610 mm outside diameter, and the designer should consult with the friction material manufacturer to determine which stock size would best fit his requirements.

Table 7.4 Friction, and allowable operating conditions for various clutch facing materials

Operating dry	μ	Temperature, °C		Working pressure kN/m^2	Power rating W/mm^2
		Maximum	Continuous		
LIGHT DUTY					
Woven	0.35–0.40	250	150	175–520	0.3–0.6
Millboard	0.40	250	150	175–700	0.3–0.6
MEDIUM DUTY					
Asbestos/non asbestos wound tape and yarn	0.40	350	200	175–700	0.3–1.2
Moulded	0.35	350	200	175–700	0.6–1.2
HEAVY DUTY					
Sintered*	0.36/0.30‡	500	300	350–2800	1.7
Cermet†	0.40			700–1400	4.0
OIL IMMERSED					
Paper	0.11			700–1750	2.3
Woven	0.08			700–1750	1.8
Moulded	0.04			700–1750	0.6
Moulded (grooved)	0.06				
Sintered	0.10/0.05‡			700–4200	2.3
Sintered (grooved)	0.10/0.06‡			700–4200	2.3

* Facings are sintered on to core plates or backing plates.
† Supplied as buttons in steel cups.
‡ First figure static, second dynamic coefficient.

Friction clutches 7

Table 7.5 Typical physical and mechanical properties of clutch facings

	Resin-based materials	Sintered metals
Thermal conductivity	0.80 W/m°C	16 W/m°C
Specific heat	1.25 kJ/kg°C	0.42 kJ/kg°C
Thermal expansion	0.5×10^{-4}/°C	0.13×10^{-4}/°C
Specific gravity	1.6 for woven/wound 2.8 for moulded	6.0
Brinell hardness number	6–15	13
	MN/m^2	MN/m^2
Ultimate tensile strength	20	45
Ultimate shear strength	12	35
Ultimate compressive strength	100	150

If the maximum temperatures in the tables are exceeded the μ may fall badly. Any increase in temperature will increase the wear rate. Figure 7.20 shows how the wear of resin-based materials typically increases with temperature, taking the wear at 100°C as unity.

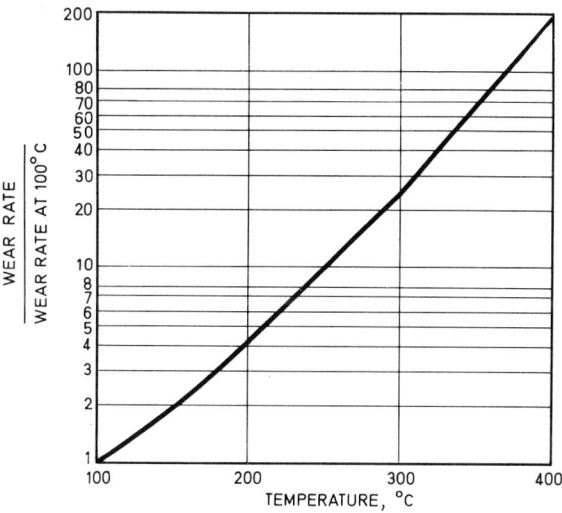

Figure 7.20 *The effect of temperature on the wear rate of clutch facing materials*

8 Brakes

A brake has to develop the required torque in a stable and controlled manner, and must not reach temperatures high enough to impair its performance, or damage its components. There are three main types of brake: band brakes, drum brakes and disc brakes.

A brake is characterised by its Brake Factor which is defined as, for band and drum brakes, as the frictional force at the drum radius divided by the actuating force, provided it is the same on both shoes (for drum brakes). For disc brakes the brake factor = 2μ where μ is the coefficient of friction.

BAND BRAKES

A flexible steel band lined with friction material is tightened against a rotating drum. Because of its self-servo action a band brake can be made very powerful. Positive self servo occurs when the frictional force augments the actuating force so increasing the torque, that is, the brake has a high brake factor. The brake factor increases rapidly with μ and the angle of wrap δ, in the case of simple band brakes, as shown in Table 8.1.

Too much self-servo makes the brake unstable and likely to grab and judder (it is usual to work with $\delta = 270°$ and $\mu = 0.3$–0.4).

The relationship between drum diameter and torque capacity for band brakes of conventional proportions is shown in Figure 8.2.

If the drum rotates in the opposite sense to the actuating force (negative servo) the brake factor is very small, but by suitable design of the actuating mechanism the brake can be made equally effective for both directions of rotation.

As band brakes require small actuating forces they are generally suitable for manual operation, particularly when they are used only intermittently.

In band brakes, the width of the rubbing path is typically one fifth of the diameter.

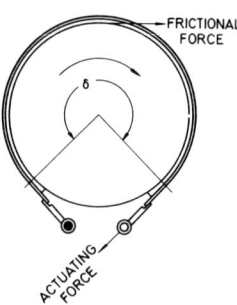

Table 8.1 Brake factor for different δ and μ

δ degrees	μ				
	0.1	0.2	0.3	0.4	0.5
210	0.44	1.08	2.00	3.33	5.25
240	0.52	1.31	2.51	4.34	7.13
270	0.60	1.57	3.11	5.59	9.55
300	0.69	1.85	3.81	7.12	12.72
330	0.78	2.16	4.63	9.00	16.82
360	0.87	2.51	5.59	11.34	22.16

Table 8.2 The various types of band brake

Type	Brake factor ($\mu = 0.3$)	Uses	Type
Simple	3.11 $\delta = 270°$	Winches, hoists, excavators, tractors, etc.	
Reversible	High; depends on lever ratio	As above but where the brake has to be equally effective in either direction	

Brakes 8

Table 8.2 (continued)

Type	Brake factor ($\mu = 0.3$)	Uses	Type
Screw-operated (reversible)	1.81 $\delta = 155°$ for each lining	Winches, reeling brakes for marine applications	
Differential	Very high; depends on lever ratio	Parking or holding brake only, as the brake is unstable	
Double wound	15.9 $\delta = 540°$	In rugged installations such as oil-well equipment where operation is manual and precise control not required	

RIGID SHOE (DRUM) BRAKES

Shoes are lined with friction material, usually over an arc of 90–110 deg. Shoes may be leading (positive self-servo) or trailing (negative self-servo). By suitable combination of shoes, or by altering the geometry of the brake, the amount of servo required for the actuating force available, and the amount of stability required, can be obtained.

On most drum brakes the actuating force on each shoe is the same and the brake factor B is defined as:

$$B = \frac{\text{Total frictional force at drum radius}}{\text{Actuating force}} = \frac{P + P'}{F}$$

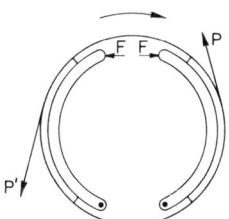

and some values of B are shown plotted against μ for different brakes in Figure 8.1

Knowing the torque, type of brake, its diameter, and the μ of the linings, the actuating force required can be obtained from Figure 8.1 (μ would normally lie between 0.3 and 0.4—a lining with the lower μ would give a more stable brake and would last longer, but a larger actuating force would be required).

Alternatively, knowing the torque required, the type of brake, and the μ of the linings, the actuating force for a given diameter brake, or the diameter of the brake for a given actuating force, can be determined from Figure 8.1.

The rubbing path area must be adequate to keep temperatures within limits so that performance is not impaired and the friction material has a good life. Table 8.5 shows the area required for different types of duty.

Smaller, wider drums of sufficient thermal capacity reduce bending stresses and give more uniform pressure on the linings, and reduce the moment of inertia of the drum. Drums for internal expanding brakes are generally stiffened with external circumferential ribs to avoid 'bell-mouthing' but can 'barrel' if ribs constrain ends but not the centre of the rubbing path. These fins also improve cooling; axial fins may develop high thermal stresses.

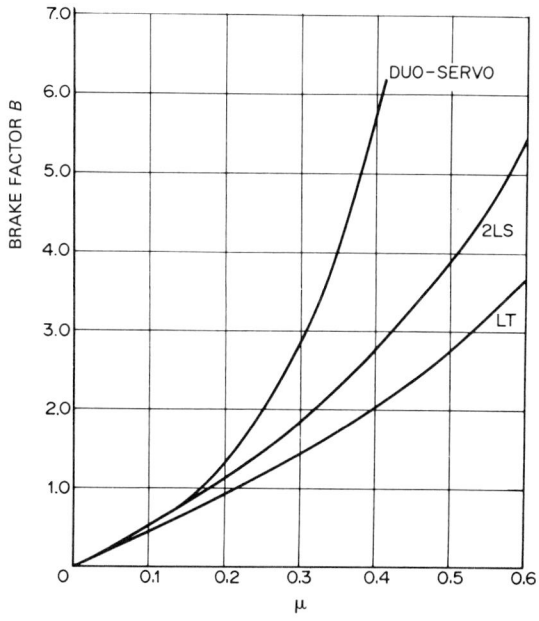

Figure 8.1 *Typical variation of Brake Factor B with μ for a LT (leading-trailing shoe), a 2LS (two leading shoe), and a duo-servo drum brake having linings of arc length 100°*

8 Brakes

Table 8.3 The various types of drum brake

Type	Brake factor ($\mu = 0.35$)	Uses	Type
Post brake (external contracting, leading–trailing shoe)	1.7	Most widely used type of industrial brake. Used on mills, colliery winders, lift equipment, etc.	
Internal expanding, leading–trailing shoe	1.7	Mainly used for automotive purposes	
Internal expanding, two leading shoe	2.3	Mainly used for automotive purposes	
Duo-servo (shoes linked together)	4.0	Mainly used for automotive purposes	

The torque capacity of drum brakes of various sizes in shown in Figure 8.2

DISC BRAKES

A caliper carrying friction pads straddles the rotating disc and the pads are forced against the disc to apply the brake. The brake factor is low and equal to 2μ, but the disc brake is stable and less affected by high temperatures than the band or drum brake. The diameter of the disc and the area of friction material can be obtained from the torque curves of Figure 8.2 or from Table 8.5. The area of the pads of a spot-type disc brake is about one tenth of the area of the rubbing path. The actuating force required to act on each paid is $F = T/\mu D$, where T is the torque, and D the diameter to the midline of the pads.

Table 8.4 The various types of disc brake

Type	Brake factor ($\mu = 0.3$)	Uses	Type
Spot type	0.70	Cars and light commercial vehicles, reel tension brakes for paper mills, cable winding equipment, steel strip mills, forge brakes, marine shaft brakes, wind generators, etc.	
Plate type	0.70 (two rubbing faces)	Machine tools, and other machinery hydro-electric turbines agricultural tractors	

Brakes 8

Table 8.5 Required brake areas for various duties

Duty	Typical application	Area/Power mm²/w	
		Band and drum brakes	Disc brake (Spot type)
Infrequent (time to cool to ambient temperature between applications)	Emergency brakes; safety brakes	0.5	0.17
Intermittent	General-duty applications, winding engines, cranes, lifts	1.7	0.43
Heavy	Excavators, presses, drop stamps, haulage gear	3.4–4.2	0.86

Torque capacity

The torque capacity of band, drum and disc brakes of conventional proportions for medium-duty applications is shown in Figure 8.2

For light-duty work multiply figures on the torque axis by 3 for band and drum brakes, by 2.5 for disc brakes, for brakes of a given diameter.

For heavy-duty work divide figures on the torque axis by 2 for band, drum and disc brakes.

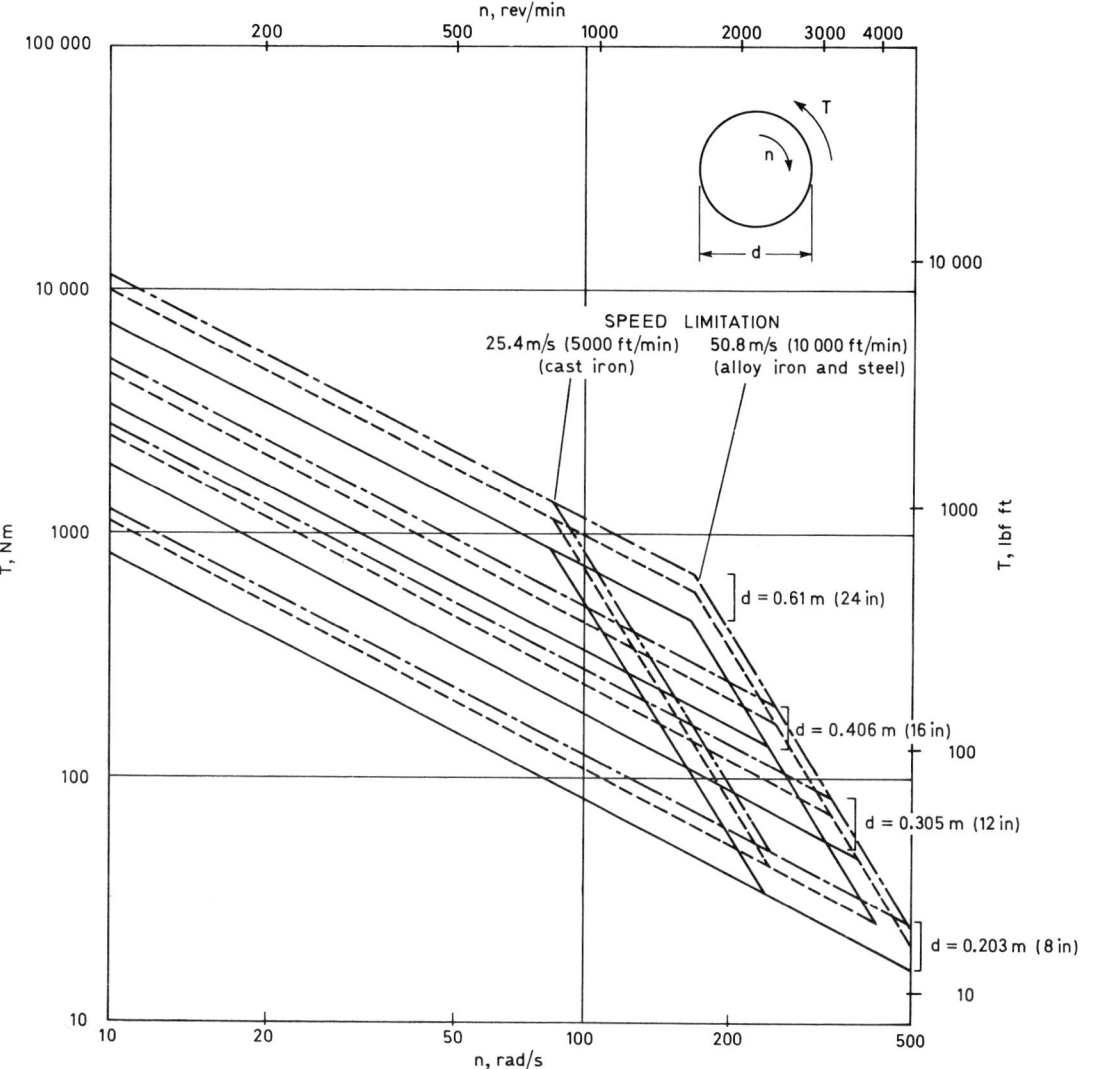

Figure 8.2 The torque capacity of band, drum and disc brakes of conventional proportions for medium-duty applications

Band brakes—The width of lining is $\frac{1}{5}$ the diameter ————
  (Rating: 1.7 mm²/W, 2 in²/hp)

Drum brakes—The width of lining is $\frac{1}{5}$ the diameter — — —
  (Rating: 1.7 mm²/W, 2 in²/hp)

Disc brakes—Length of pad $\frac{1}{10}$ mean diameter, width $\frac{3}{4}$ length — - —
  (Rating: 6.8 mm²/W, 8 in²/hp)

8 Brakes

BRAKE SELECTION

Table 8.6 Brake selection for special performance requirements

Type of brake	Maximum operating temperatures	Brake factor	Stability	Standard parts available	Setting up	Type
Positive servo band brakes	Low	Very high	Very low	Some	Straightforward but watch excessive expansion in narrow tracks	
Drum brakes with two leading shoes	Low for external shoes, higher for internal shoes	High	Low	Some	Watch drum ovality and run-out and drum stiffness	
Drum brakes with one leading and one trailing shoe	Low for external shoes, higher for internal shoes	Moderate	Moderate	Some	Watch drum ovality and run-out and drum stiffness	
Duo-servo brakes	Low	Very high	Low	Some	Watch drum ovality and run-out	
Disc brakes	High	Low	High	Some	Watch disc run-out thick/thin variation round disc, disc stiffness, disc coning.	

Brakes 8

Table 8.7 Brake selection for special environmental conditions

Type of brake	High temperature	Low temperature	Wet and humid	Dirt and dust	External vibration	Type
Band	Good up to temperature limit of friction material	Good but avoid ice formation	Unstable but still effective	Good	Good	
Drum	Good up to temperature limit of friction material	Good but avoid ice formation	Unstable if humid. If very wet complete loss of performance on internal brakes but external brakes are more effective	Very good if sealed	Good	
Disc	Good up to temperature limit of friction material	Good but avoid ice formation	Good	Poor, should be shielded	Good	
General comments	Watch effect of thermal expansion on clearances		Watch corrosion and 'stiction' after prolonged parking in high humidity conditions			

Table 8.8 Methods of actuation

	Advantages	Disadvantages	Points to watch	Uses
Mechanical	Robust. Simple. Manual operation gives good control	Large leverage needed	Frictional losses at pins and pivots	Band brakes Drum brakes Disc brakes
Pneumatic	Large forces available	Compressed air supply needed. Brake chambers may be bulky. Slow response	Length of stroke (particularly if diaphragm type)	Band brakes Drum brakes Disc brakes
Hydraulic	Compact. Large forces available. Quick response and good control	Special fluid needed. Temperatures must not be high enough to vaporise fluid	Seals	Band brakes Drum brakes Disc brakes (spot type)
Electrical	Suitable for automatic control, quick response	On/off operation	Air gap	Band brakes Drum brakes Disc brakes (spot and plate types)

Note: Many brakes fail-safe; powerful springs apply the brake which is held off by one of the above means. Reduction in, for example, hydraulic pressure applies the brake.

8 Brakes

FRICTION MATERIALS

A very wide range of friction materials is available, and in many cases materials have been developed for specific applications. The friction material manufacturer should therefore be consulted at an early stage in the design of the brake and should also be consulted concerning stock sizes – standard sizes are much cheaper than non-standard and are likely to be immediately available. The non-asbestos lining materials are normally made in flexible rolls in standard lengths, (e.g. 4 m) and widths (330 mm) and various thicknesses. Linings of the required sizes are slit from the standard sheets. These linings are bonded to the shoes and, by increasing the temperature and time of bonding, the linings can be made more rigid, and able to withstand higher and higher duties.

Industrial disc pads are generally based on automobile and CV pad types. They may be classified as organic-non-asbestos, low steel, or semi-metallic pads. They are based on thermosetting polymers reinforced by inorganic (e.g. glass) or organic fibres, 10–15%, or 50% by weight of steel fibre; and suitable fillers are added to give the pads the required tribological properties. The organic non-asbestos pads are suitable for lighter duties, and the greater the amount of steel fibre the higher the temperature the pads can withstand. The non-metallics tend to give less squeal and groan and cold judder, and less lining and rotor wear at low temperatures; steel fibres give a more stable μ and better high temperature lining life, but they can cause corrosion problems and they allow more heat to pass into the brake assembly instead of into the disc.

Data for typical materials are shown in Tables 8.9, 8.10 and 8.11. These figures are meant as a guide only; materials vary from manufacturer to manufacturer, and any one manufacturer may make up a number of different materials of the one type which may vary somewhat in properties.

Table 8.9 Material types and applications

Type	Manufacture	Typical dimensions	Uses
LININGS*			
Woven cotton	Closely woven belt of fabric is impregnated with resins which are then polymerised	As rolls; thickness 3.2–25.4 mm width up to 304.8 mm and lengths up to 15.2 m	Industrial drum brakes, minewinding equipment, cranes, lifts
Woven asbestos	Open woven belt of fabric is impregnated with resins which are then polymerised. May contain wire to scour the surface	As radiused linings thickness 3.2–12.7 mm width up to 203 mm, minimum radius 76 mm, maximum arc 160°	Industrial band and drum brakes, cranes, lifts, excavators, winches, concrete mixers. Mine equipment
Non-asbestos flexible semi-flexible rigid	Steel, glass or inorganic fibre and friction modifiers mixed with thermosetting polymer and mixture heated under pressure	Linings: thicknesses up to 35 mm Maximum radius about 15–30 times thickness depending upon flexibility	Industrial drum brakes Heavy-duty drum brakes— excavators, tractors, presses
PADS			
Resin-based	Similar to linings but choice of resin not as restricted as flexibility not required	In pads up to 25.4 mm in thickness or on backplate to fit proprietary calipers	Heavy-duty brakes and clutches, press brakes, earth-moving equipment
Sintered metal	Iron and/or copper powders mixed with friction modifiers and the whole sintered		Heavy-duty brakes and clutches, press brakes, earth-moving equipment
Cermets	Similar to sintered metal pad, but large proportion of ceramic material present	Supplied in buttons, cups	As above

* Many lining materials supplied as large pads can be bolted, or riveted, using brass rivets, to the band or shoe; the pads can be moved along the band or shoe as wear occurs and so maximum life obtained from the friction material despite uneven wear along its length. Alternatively, and particularly with weaker materials, the friction material can be bonded to the metal carrier using proprietary adhesives and techniques (contact the manufacturer). On safety-critical applications the friction material should be attached by both bonding and riveting.

Brakes 8

Table 8.10 Performance and allowable operating conditions for various materials

Materials	μ	Temperatures, °C		Working pressures kN/m²	Maximum pressure MN/m²
		Maximum	Maximum operating		
LININGS					
Woven cotton	0.50	150	100	70–700	1.5
Woven asbestos	0.40	250	125	70–700	2.1
Non-asbestos					
light duty (flexible)	0.38	350	175	70–700	2.1
medium duty (semi-flexible)	0.35	400	200	70–700	2.8
heavy duty (rigid)	0.35	500	225	70–700	3.8
PADS					
Resin based	0.32	650	300	350–1750	5.5
Sintered	0.30	650	300	350–3500	5.5
Cermet	0.32	800	400	350–1050	6.9

Table 8.11 Typical mechanical properties

Materials	Ultimate strength			Rivet-holding capacity	Specific Gravity
	Tensile MN/m²	Shear MN/m²	Compressive MN/m²	MN/m²	
LINING					
Woven cotton	20.7	12.4	96.5	69	1.0
Woven asbestos	24.1	13.8	103.4	83	1.5–2.0
Non-asbestos					
light duty	8.2	8.2	41.3	103	1.7
medium duty	10.3	8.2	96.5	152	1.7
heavy duty	13.8	9.0	103.4	172	2.0
PAD					
Resin-based	—	9.0	103.4	—	2.0
Sintered metals	48.2	68.9	151.6	—	6.0

Mating surfaces

Woven cotton or asbestos linings, and those with steel and inorganic or organic fibre reinforcement, should run against fine-grained pearlitic cast-iron or alloy cast-iron of Brinell Hardness 180–240 or steel cold-rolled or forged with a Brinell Hardness greater than 200. The surface should be fine-turned or ground to a finish of at least 2.5 μm CLA. Cast steel and non-ferrous materials are not recommended. Some friction materials are very sensitive to trace amounts of titanium (and some other elements) in the cast iron rotor, and these trace elements can considerably reduce μ, though they also tend to increase the life of the friction material.

Sintered metals should run against fine-grained pearlitic cast iron or alloy irons, Brinell Hardness 180–250. High carbon steel such as EN6 for moderately loaded, and EN42 for heavy-duty thin counterplates in multidisc clutches. Minimum Brinell Hardness 200 for heavy duty. The surface finish should be 0.9–1.5 μm CLA.

Cermets should run against similar cast irons with Brinell Hardness greater than 200. High carbon steels with a hardness between 200 and 300 are acceptable. The surface finish should again be 0.9–1.5 μm CLA.

9 Screws

Screws are used as linear actuators or jacks and can generate substantial axial forces. They can operate with an external drive to either the screw or the nut, and the driving system often incorporates a worm gear in order to obtain a high reduction ratio.

TYPES OF SCREW

Plain screws

In these screws the load is transmitted by direct rubbing contact between the screw and the nut.

These are the simplest and inherently the most robust. The thread section may be of a square profile or more commonly is of the acme type with a trapezoidal cross section.

Their operating friction is relatively high but on larger diameter screws can be reduced to very low levels by incorporating hydrostatic pads into the operating surfaces of the nut. This is usually only justified economically in special screws such as the roll adjustment screws on large rolling mills.

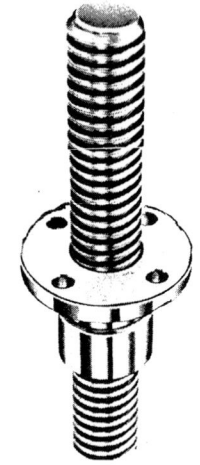

Ball screws

In these screws the load is transmitted by close packed balls, rolling between the grooves of the screw and the nut.

These provide the lowest friction and are used particularly for positioning screws in automatically controlled machines. The nuts need to incorporate a system for re-circulating the balls. The load capacity is less than in other types of screw and is limited by the contact stresses between the balls and the screw.

Planetary roller screws

In these screws a number of rollers are positioned between the screw and the nut and rotate between them, around the screw, with a planetary motion.

Those with the highest load capacity have helical threads on the rollers and nut, matching the pitch of the screw. The whole space between the screw and the nut can be packed with rollers but these need to have synchronised rotation by a gear drive to ensure that they retain their axial position.

Alternative types are available in which the rollers and nut have simple parallel ribs matching the pitch of the screw. The screw however has to be multistart because the number of rollers that can be fitted equals the number of starts on the thread. Also the nut cannot be used as the driver if synchronised external movement is required, because of the possibility of slip between the rollers and the nut. In these cases the screw or the planetary roller carrier has to be driven, but not the nut.

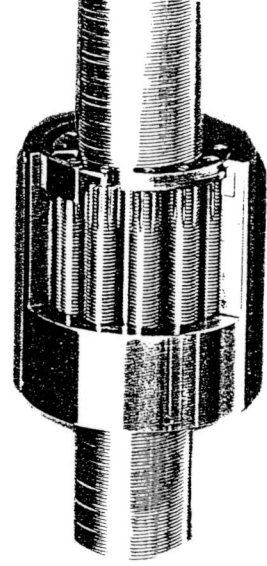

Screws 9

PERFORMANCE

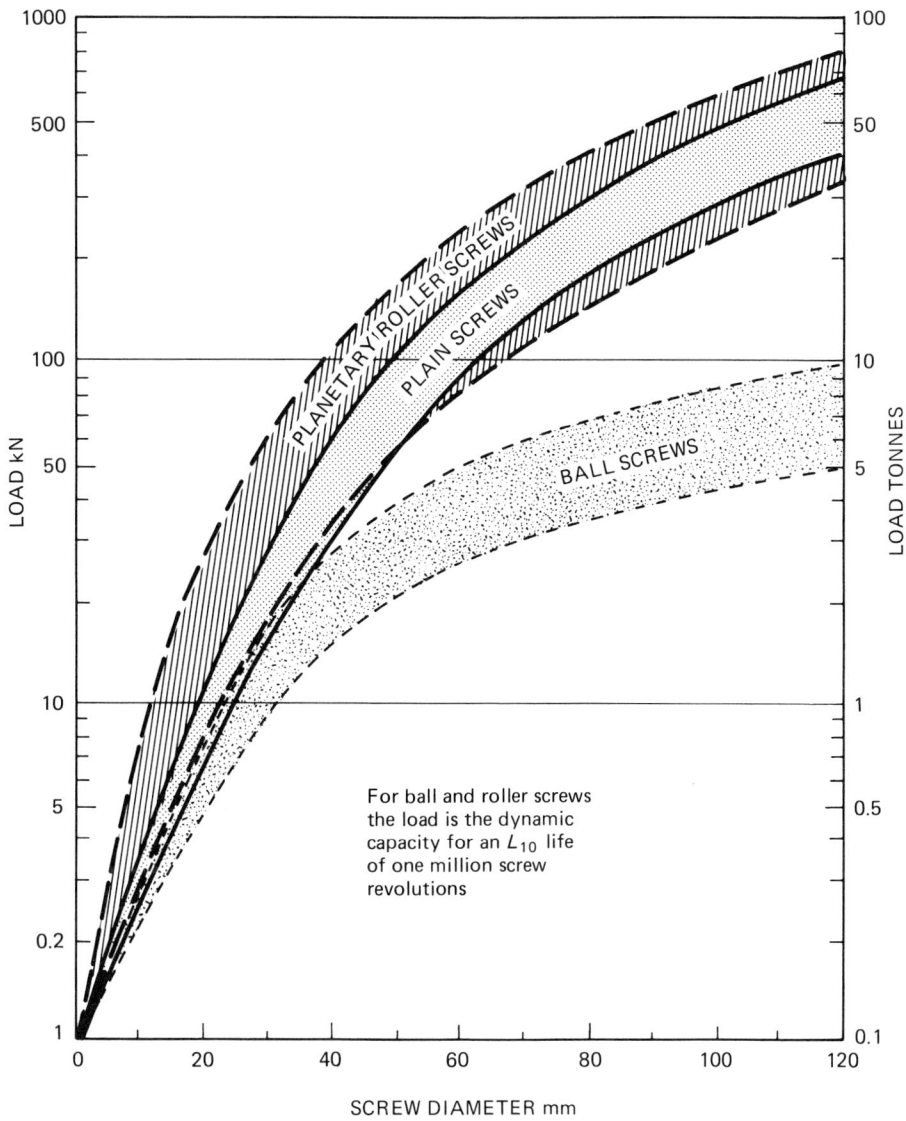

Figure 9.1 The axial load capacity of various types of screw

Mechanical efficiencies of screws

Plain screws   30%–50% approximately, with larger diameter screws tending to have the lower values.
Roller screws   65%–85%
Ball screws   75%–90%

9 Screws

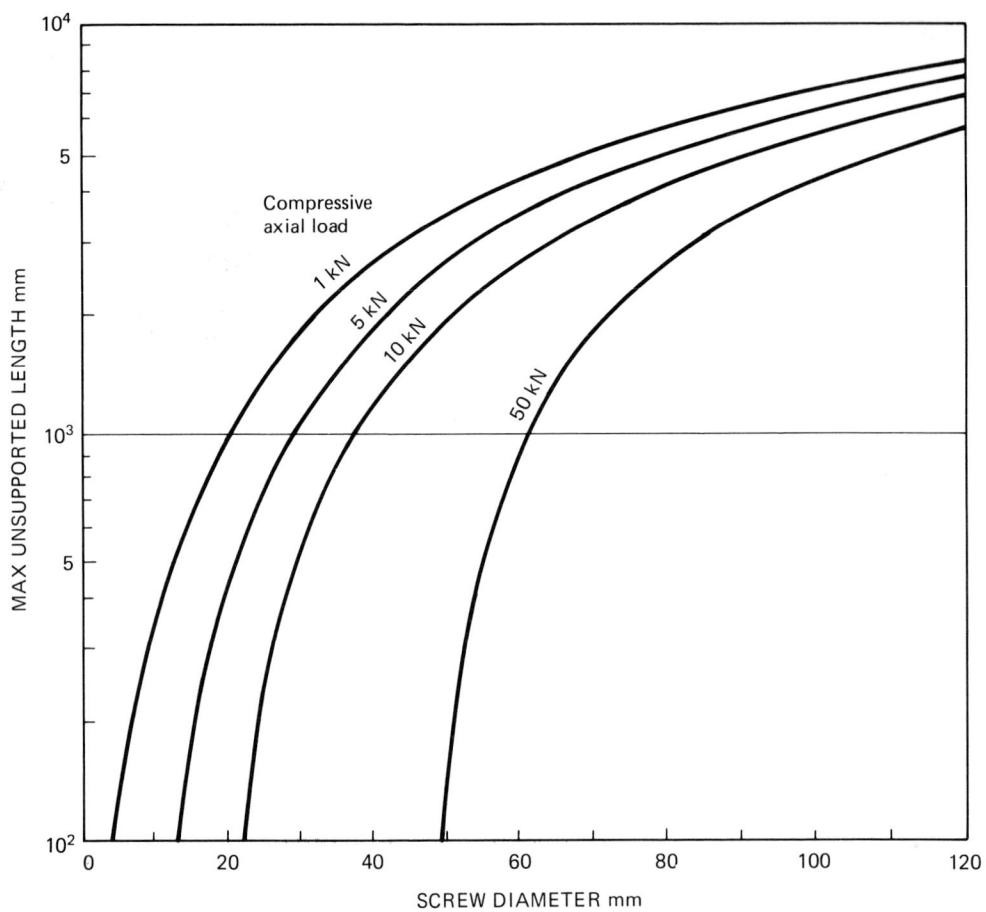

Figure 9.2 The maximum unsupported length of screw, with one end free, to avoid buckling

Installation

The performance of all screws will be reduced when subject to misalignment and sideways loads. Ball screws are particularly sensitive to these effects.

All screws require lubrication, either by regular greasing or by operation within an enclosure with oil or fluid grease.

For precision installations the screws and nuts need protection from external contamination and flexible convoluted gaiters are commonly used for their protection.

Cams and followers   10

COMMON MODES OF FAILURE

Three main forms of cam and tappet failures occur. These are pitting, polish wear and scuffing. Failure may occur on either the cam or tappet, often in differing degree on both.

Pitting

This is the failure of a surface, manifested initially by the breaking-out of small roughly triangular portions of the material surface. This failure is primarily due to high stresses causing fatigue failure to start at a point below the surface where the highest combined stresses occur. After initiation a crack propagates to the surface and it may be that the subsequent failure mechanism is that the crack then becomes filled with lubricant, which helps to lever out a triangular portion of material.

Heavily loaded surfaces will continue to pit with increasing severity with time. Figure 10.1 shows some pitted cam followers.

Polish wear

This is the general attrition of the contacting surfaces. When conditions are right this will be small, but occasionally very rapid wear can occur, particularly with chilled and hardened cast iron flat-faced tappets. Often a casual look will suggest that the surfaces are brightly polished and in good condition but dimensional checks reveal that considerable wear has occurred. Polish wear appears to be an intermediate case between pitting and scuffing assisted by some form of chemical action involving the oil – certainly surfaces which develop a bloom after running do not normally give 'polish wear'.

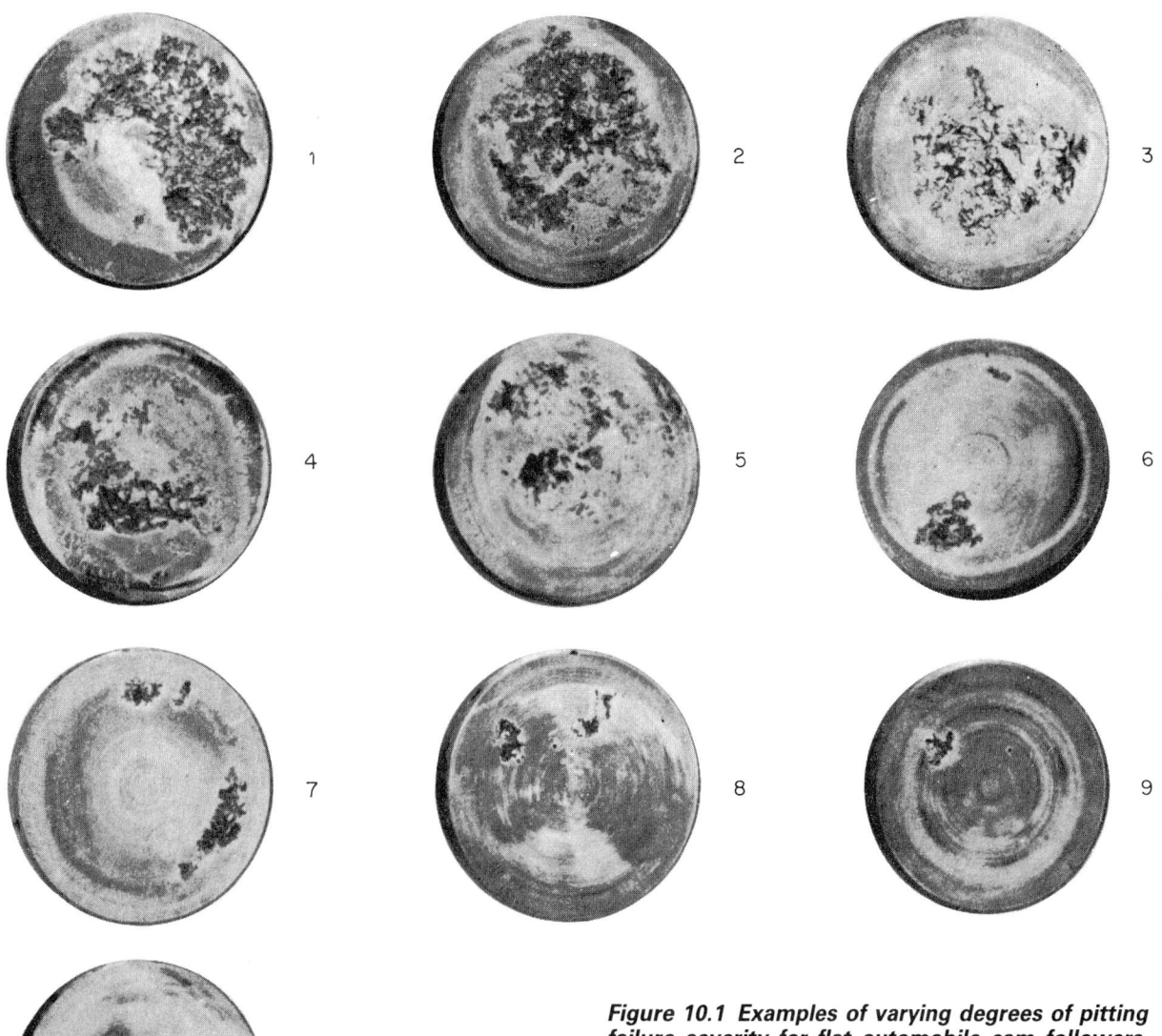

Figure 10.1 Examples of varying degrees of pitting failure severity for flat automobile cam followers. Numbers indicate awards for ratings for lack of damage during oil standardisation tests (courtesy Orobis Ltd.)

10 Cams and followers

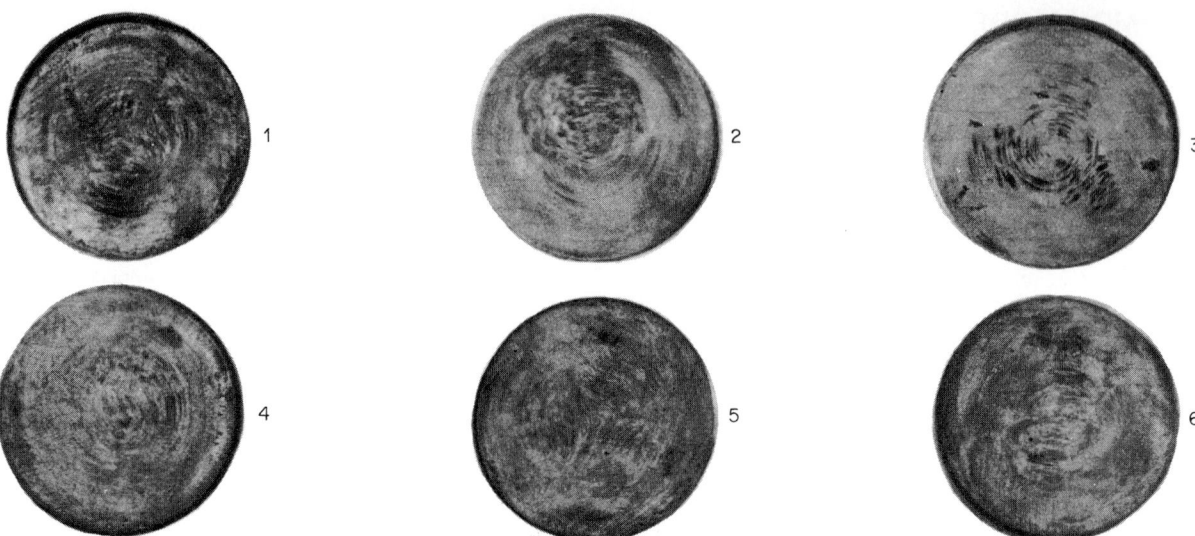

Figure 10.2 Examples of varying severity of cam follower, scuffing for flat automotive tappets (courtesy: Orobis Ltd.)

Scuffing

This is the local welding together of two heavily loaded surfaces, particularly when a high degree of relative sliding occurs under poor lubrication conditions, followed by the tearing apart of the welded material. It is particularly likely to start from high spots, due to poor surface finish, during early running of new parts.

CHECKING THE TRIBOLOGICAL DESIGN

It is usual to assess cam/tappet designs on the basis of the maximum contact stress between the contacting cam and tappet, with some consideration of the relative sliding velocity. This requires the determination of the loads acting between the cam and tappet throughout the lift period (at various speeds if the mechanism operates over a speed range), the instantaneous radius of curvature for the cam throughout the lift period, and the cam follower radius of curvature. Figure 10.3 shows the relationship between these various quantities for a typical automotive cam. In addition it is possible to assess the quality of lubrication at the cam/tappet interface by calculating the elastohydrodynamic (EHL) film thickness and relating this to the surface roughness of the components. An approximate method for the calculation is given later in this section.

Calculation of the instantaneous radius of curvature of the cam

Where the cam is made up of geometric arcs and tangents the appropriate values for the radii of curvature can be read from the drawing. Many cams are now generated from lift ordinates computed from a mathematical law incorporating the desired characteristics, so it is necessary to calculate the instantaneous radius of cam curvature around the profile. At any cam angle the instantaneous radius of curvature at that angle is given by the following:

For flat followers (tappets)

$$R_c = R_{base} + y + 3282.8\, y''$$

where R_{base} = base circle radius in mm
y = cam lift at desired angle in mm
y'' = cam acceleration at chosen angle in mm/deg^2
R_c = radius of curvature in mm

For curved followers

$$R_c = \left\{ \frac{[(R_b + R_F + y)^2 + V^2]^{3/2}}{(R_b + R_F + y)^2 + 2V^2 - (R_b + R_F + y)A} \right\} - R_F$$

where R_b = cam base circle radius in mm
R_F = follower radius in mm
y = cam lift at chosen angle in mm
V = follower velocity at chosen angle in mm/rad
  = 57.29 × velocity in mm/deg
A = follower acceleration mm/rad^2
  = 3282.8 × acceleration (mm/deg^2)

The value for R_c will be positive for a convex cam flank and negative for a concave (i.e. hollow) flank.

Cams and followers 10

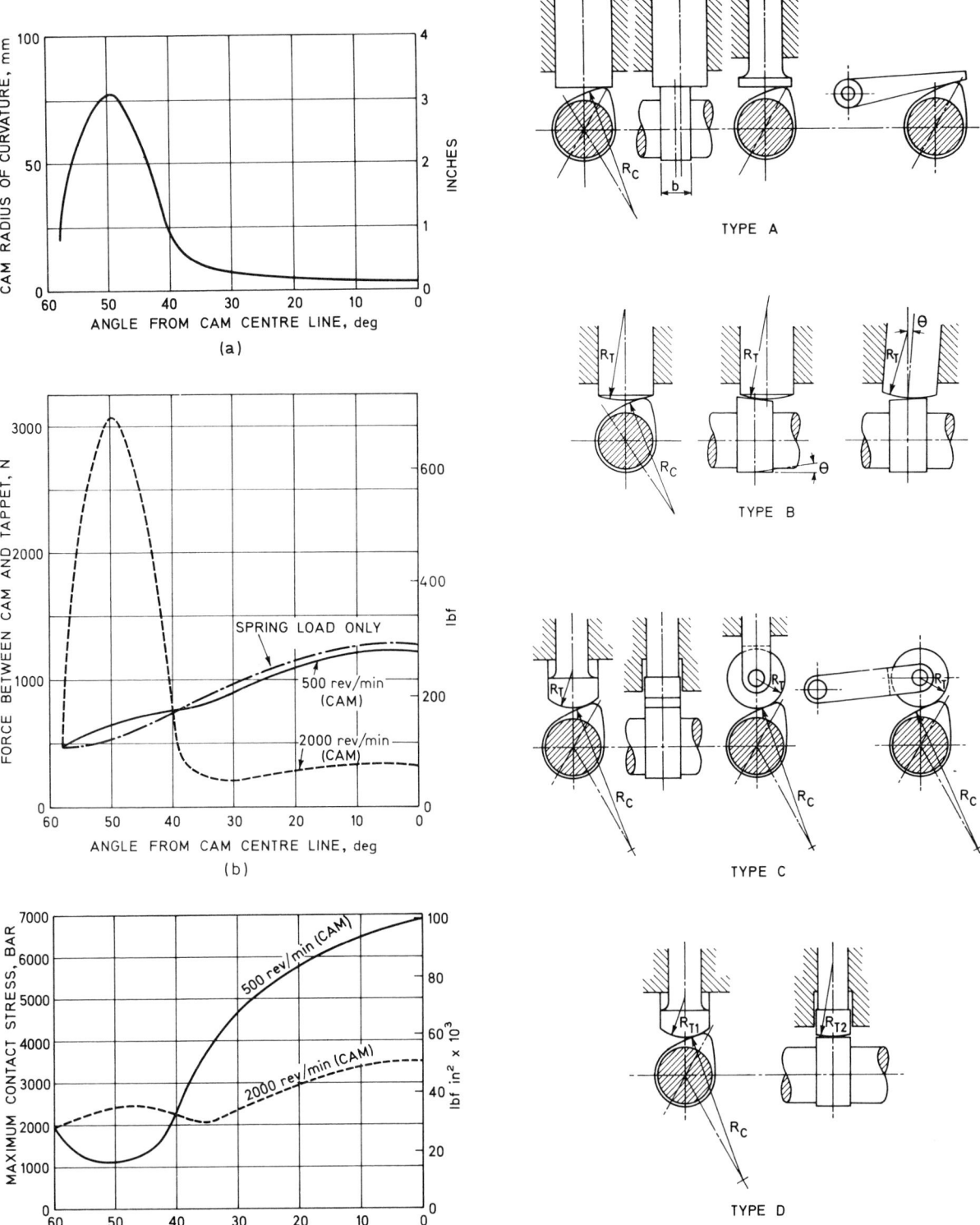

Figure 10.3 **Typical variation for an automotive cam of:** (a) instantaneous radius of curvature; (b) cam/tappet force; (c) maximum contact stress

Figure 10.4 **Classification of cams and tappets for determination of contact stresses.** Type A: flat follower faces. Type B: spherical faced tappets. Type C: curved and roller followers with flat transverse faces. Type D: curved tappets with transverse radius of curvature

10 Cams and followers

Calculation of contact (Hertzian) stress

It is now necessary to calculate the Hertzian stresses between the cam and tappet. Most tappets and cams can be classified into one of the forms shown in Figure 10.4. The appropriate formulae for the Hertzian stress are listed below.

The following symbols and units are used:

W = load between cam and tappet (N)

b = width of cam (mm)

R_c = cam radius of curvature at point under consideration (mm)

R_T = tappet radius curvature (mm)

R_{T1} = tappet radius of curvature in plane of cam (mm)

R_{T2} = tappet radius of curvature at right angles to plane of cam (mm)

f_{max} = peak Hertzian stress at point under consideration (N/mm^2)

Type A: Flat tappet face on cam

$$f_{max} = K \left[\frac{W}{R_c \cdot b} \right]^{1/2}$$

Material combination	K
Steel on steel	188
Steel on cast iron	168
Cast iron on cast iron	153

The centre line of the tappet is often displaced slightly axially from the centre line of the cam to promote rotation of the tappet about its axis. This improves scuffing resistance but is considered by some to slightly reduce pitting resistance.

Type B: Spherical faced tappet

Since the theoretical line contact of Type A tappets on the cam is often not achieved, due to dimensional inaccuracies including asymmetric deflection of the cam on its shaft, edge loading occurs. To avoid this a large spherical radius is often used for the tappet face. Automotive engines use a spherical radius of between 760 to 2540 mm (30 to 100 in). To promote tappet rotation the tappet centre line is displaced slightly from the axial centre line of the cam and the cam face tapered (10–14 min of arc with 760 mm tappet radius and 4–7 min for 1500–2540mm tappet radius). Alternatively the longitudinal tappet axis is tilted by a corresponding amount to the camshaft axis. The theoretical point contact extends into an elongated ellipse under load to give a better contact zone than with the nominally flat face.

$$f_{max} = X \cdot K \left[\frac{2}{R_T} + \frac{1}{R_c} \right]^{2/3} \cdot W^{1/3}$$

K is obtained from Figure 10.5 after evaluating

$$\left[1 + \frac{2R_c}{R_T} \right]$$

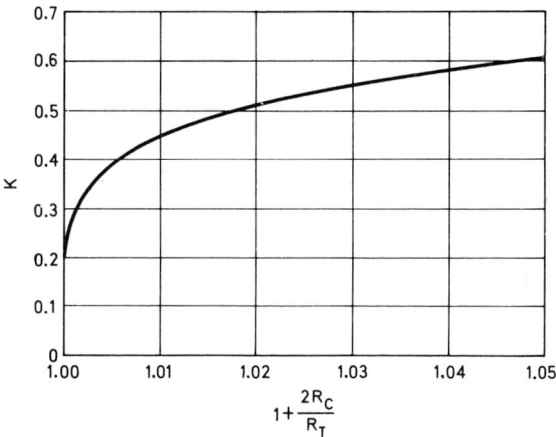

Figure 10.5 Constant for the determination of contact stresses with spherical-ended tappets

Material combination	X
Steel on steel	838
Steel on cast iron	722
Cast iron on cast iron	640

Type C: Curved and roller tappets with flat transverse face

$$f_{max} = K \left[\left(\frac{1}{R_c} + \frac{1}{R_T} \right) \frac{W}{b} \right]^{1/2}$$

Where K is the same as for type A, flat tappet face on cam.

Cams and followers   10

Type D: Curved tappet with large transverse curvature (crowning)

The large transverse radius of curvature has values similar to those used in Type B.

$$f_{max} = X \cdot K \left[\frac{1}{R_{T1}} + \frac{1}{R_{T2}} + \frac{1}{R_c} \right]^{2/3} \cdot W^{1/3}$$

X values for material combinations as for Type B.
K is obtained from Figure 10.5 after evaluating

$$\left[\frac{\frac{1}{R_{T1}} + \frac{1}{R_{T2}} + \frac{1}{R_c}}{\frac{1}{R_{T1}} - \frac{1}{R_{T2}} + \frac{1}{R_c}} \right]$$

and using scale labelled

$$\left(1 + \frac{2R_c}{R_T} \right).$$

Allowable design values for contact stress

Safe values for contact stress (Hertzian stress) are dependent on a number of factors such as the combination of materials in use; heat treatment and surface treatment; quality of lubrication.

Figure 10.6 gives allowable contact stress for iron and steel components of various hardnesses. These values can only be applied if lubrication conditions are good, and this needs to be checked using the assessment method below.

ASSESSMENT OF LUBRICATION QUALITY

Calculation of film thickness

The lubrication mechanism in non-conformal contacts such as in ball bearings, gears, and cams and followers, is Elastohydrodynamic lubrication or EHL. This mechanism can generate oil films of thicknesses up to the order of 1 μm. There is a long formula for accurately calculating the film thickness, but a simple formula is given below which gives sufficient accuracy for assessing the lubrication quality of cams and followers. This formula applies only to iron or steel components with mineral oil lubrication.

$$h = 5 \times 10^{-6} \times (\eta \, u \, R_r)^{0.5}$$

where:

h = EHL film thickness (mm)
η = lubricant viscosity at working temperature (Poise)
u = entrainment velocity (mm/s).
 – for evaluation of u, see below
R_r = relative radius of curvature (mm)
 – for flat tappets $R_r = R_c$
 – for curved tappets
  $R_r = (1/R_c + 1/R_T)^{-1}$
 – for spherical or barrelled roller tappets assume $R_T = R_{T_1}$

Evaluation of entrainment velocity u

The entrainment velocity u can vary enormously through the cam cycle, reversing in sign, and in some cases remaining close to zero for part of the cycle. This last condition leads to very thin or zero thickness films.

For roller followers, u can be taken as being approximately the surface speed of the cam. Calculation of the EHL film is only required at the cam nose and on the base circle. Roller followers usually have good lubrication conditions at the expense of high contact stress for a given size.

For plain tappets the entrainment velocity, u at any instant is the mean of the velocity of the cam surface relative to the contact point and the velocity of the follower surface relative to the contact point.

On the base circle therefore, where the contact point is stationary, u is half the cam surface speed.

At all other parts of the cycle, the contact point is moving. The entrainment velocity u can be calculated from the following equation.

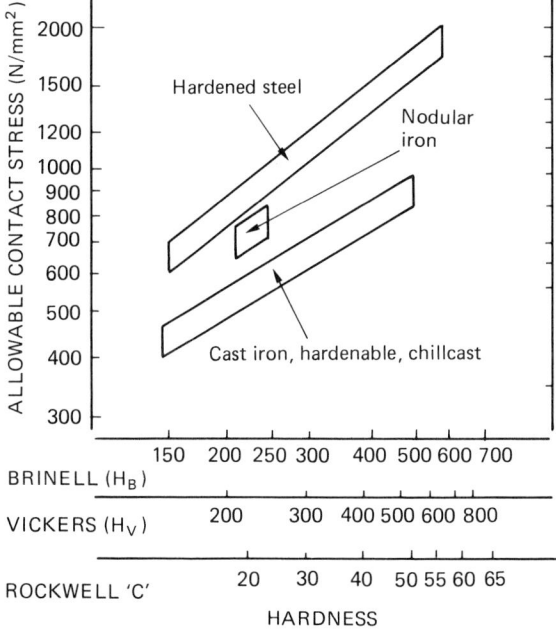

Figure 10.6 Typical allowable contact stresses under good lubrication conditions

10 Cams and followers

$$u = \omega \left[\frac{R_b}{2} + \frac{y}{2} - R_c \right]$$

ω = cam speed in rad/s
R_b = base circle radius (mm)
y = cam lift (mm)
R_c = cam radius at point of contact (mm)

This applies for flat tappets, and for curved tappets with a radius much larger than the cam radius it can be used as a reasonable approximation.

Ideally the values for u and R_r should be calculated for all points on the cycle, but as a minimum they should be calculated for the base circle and the maximum lift position.

For cams with curved sliding contact followers the equation for u is very complex. However, to check the value of u at the maximum lift position only, the following approximate formula can be used.

$$u = \omega \left[\frac{3280 y'' R_F}{(R_F + R_b + y)} + \frac{(R_b + y)}{2} \right]$$

(curved follower, max lift position only)

where

y = max cam lift (mm)
y'' = max cam acceleration at nose (mm/deg^2)
    which is a negative value
R_F = follower radius (mm)
R_b = base circle radius (mm).

Evaluation of mode of lubrication

Once a value for the film thickness has been calculated, the mode of lubrication can be determined by comparing it with the effective surface roughness of the components. The effective surface roughness is generally taken as the combined surface roughness R_{qt}, defined as

$R_{qt} = (R_{q1}^2 + R_{q2}^2)^{0.5}$

R_{q1} and R_{q2} are the RMS roughnesses of the cam and tappet respectively, typically 1.3 times the R_a (or CLA) roughness values.

If the EHL film thickness h is greater than R_{qt} then lubrication will be satisfactory.

If the EHL film thickness is less than about 0.5 R_{qt} then there will be some solid contact and boundary lubrication conditions apply. Under these circumstances, surface treatments and surface coatings to promote good running-in will be desirable, and anti-wear additives in the oil may be necessary.

Alternatively, it may be appropriate to improve surface finishes, or to change the design to an improved profile giving better EHL films, or to use roller followers which are inherently easier to lubricate.

SURFACE FINISH

Extremely good surface finishes are desirable for successful operation, as the EHL lubrication film is usually very thin.

Typical achievable values are 0.4 μm R_a for the cam, and 0.15 μm R_a for the tappets.

SURFACE TREATMENTS

Some surface treatment and heat treatment processes which can be used with cams and tappets are given below:

Phosphating	Running-in aid. Retains lubricant.
'Tufftride'	Running-in aid. Scuff resistant
'Noscuff'	Greater depth than Tufftride. Less hard.
'Sulf. B.T.'	Low distortion. Anti-scuffing
Flame hardening, Induction hardening	Can give distortion
Laser hardening	Low distortion. 0.25 to 1 mm case depth
Carburising	0.5 mm case depth typical
Nitriding	Depth 0.3 mm. Hardening and scuff resistance.
Plasma Nitriding	As nitriding, but low distortion
Sulfinuz	Good scuff resistance
Boriding	Good wear resistance

OIL AND ADDITIVES

The oil type is frequently constrained by requirements of other parts of the machine. However, for best lubrication of the cam and tappet (i.e.: thickest EHL film), the viscosity of the lubricant at the working temperature should be as high as possible. Often the best way of achieving this is to provide good cooling at the cams, by means of a copious supply of oil.

Trends in vehicle engine design such as overhead camshafts, and higher underbonnet temperatures, have led to high camshaft temperatures and low lubricant viscosities. Some cam wear problems may be partially attributed to this.

Oil additives, principally ZDDP (zinc-dialkyldithiophosphate) and similar, are used in vehicle engine oils, and are beneficial to cam and tappet wear. There is evidence that these additives can promote pitting at high temperatures, due to their chemical effects. Additives should therefore be used with care, and are certainly not an appropriate alternative to good design.

Wheels, rails and tyres 11

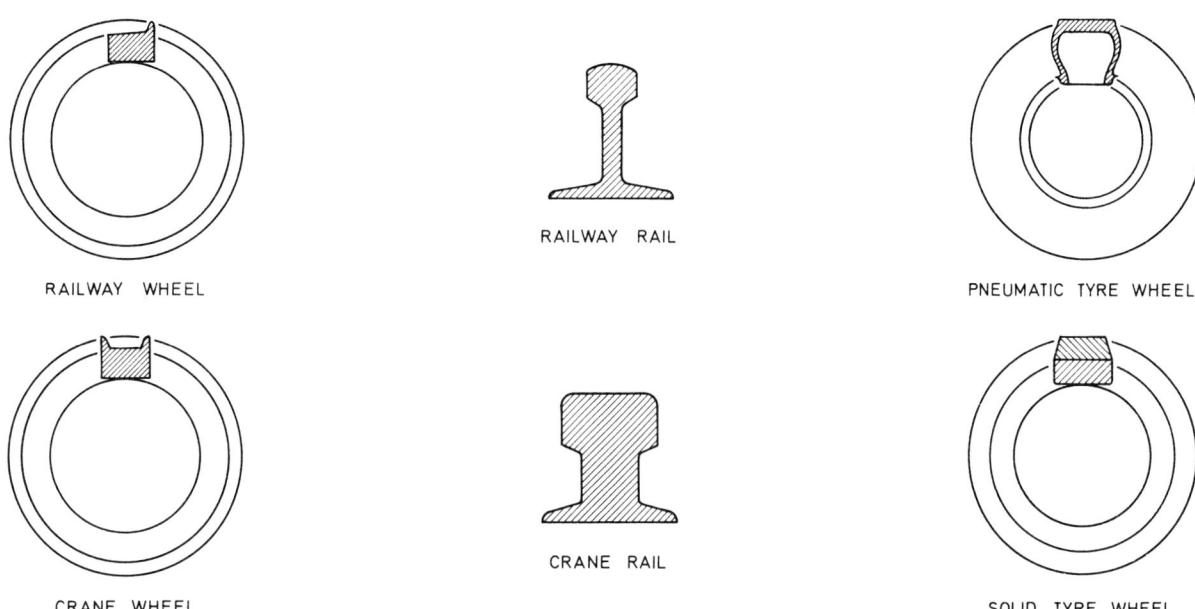

Figure 11.1 Cross sections of typical wheels and rails

Table 11.1 The effect of various factors on wear

	Load (W) lbf or N	Speed	Elevated temp.	Water	Wheel dia. in or mm	Material composition	Mating surface	Other
Steel rails (vertical wear)	See Figure 11.2. Wear $\alpha\ W^\beta$ where $\beta = \frac{1}{2}$ to 1	Little overall effect except on side or flange wear. Increased wear at high speeds counteracted by increased wear due to more acceleration, gradient climbing, and braking at low speeds	Increased corrosive wear at temperatures above 200°C, depending on the atmosphere	Increased wear due to corrosion. Hence the slope of the lines in Figure 11.2	Rail wear reduced by increasing wheel size	Increased hardness and corrosion resistance reduce wear substantially. Steel should have a fine pearlitic microstructure	Wear is increased by a rough or distorted mating surface, or by sanding the track	Atmospheric pollution increases wear substantially
Steel tyres	Similar to rail wear See Figure 11.2		As for rails	Little effect	Wear α (diameter)$^\beta$ where $\beta = 2$ to 3	Increased hardness and toughness reduce wear	Corrugated rail surfaces cause wear and noise. Curves in the track increase wear	
Pneumatic rubber tyres	See Figure 11.5	See Figure 11.5 Wear due mainly to the effect of braking and cornering. One right-angled turn equivalent to 3 stops	Maximum temperature for continuous use 120°C	Wear reduced to 20–50% of dry wear	Wear α (diameter)$^\beta$ where $\beta \leqslant 2$	Radial ply construction reduces wear by about 40%. The wear resistance of different rubber varies widely	Harshness more important than roughness in road surfaces, by a factor of about 3. See Figure 11.5	For effect of inflation pressure, see Figure 11.5
Solid tyres	Proportional to W^γ where $\gamma > 1$	Use limited to low speeds due to poor heat dissipation	Important limiting factor: dependent on tyre material	Reduced wear	Wear decreases with increasing size	Materials should be chosen to suit the conditions	Rough, sharp surfaces cause rapid wear	

11 Wheels, rails and tyres

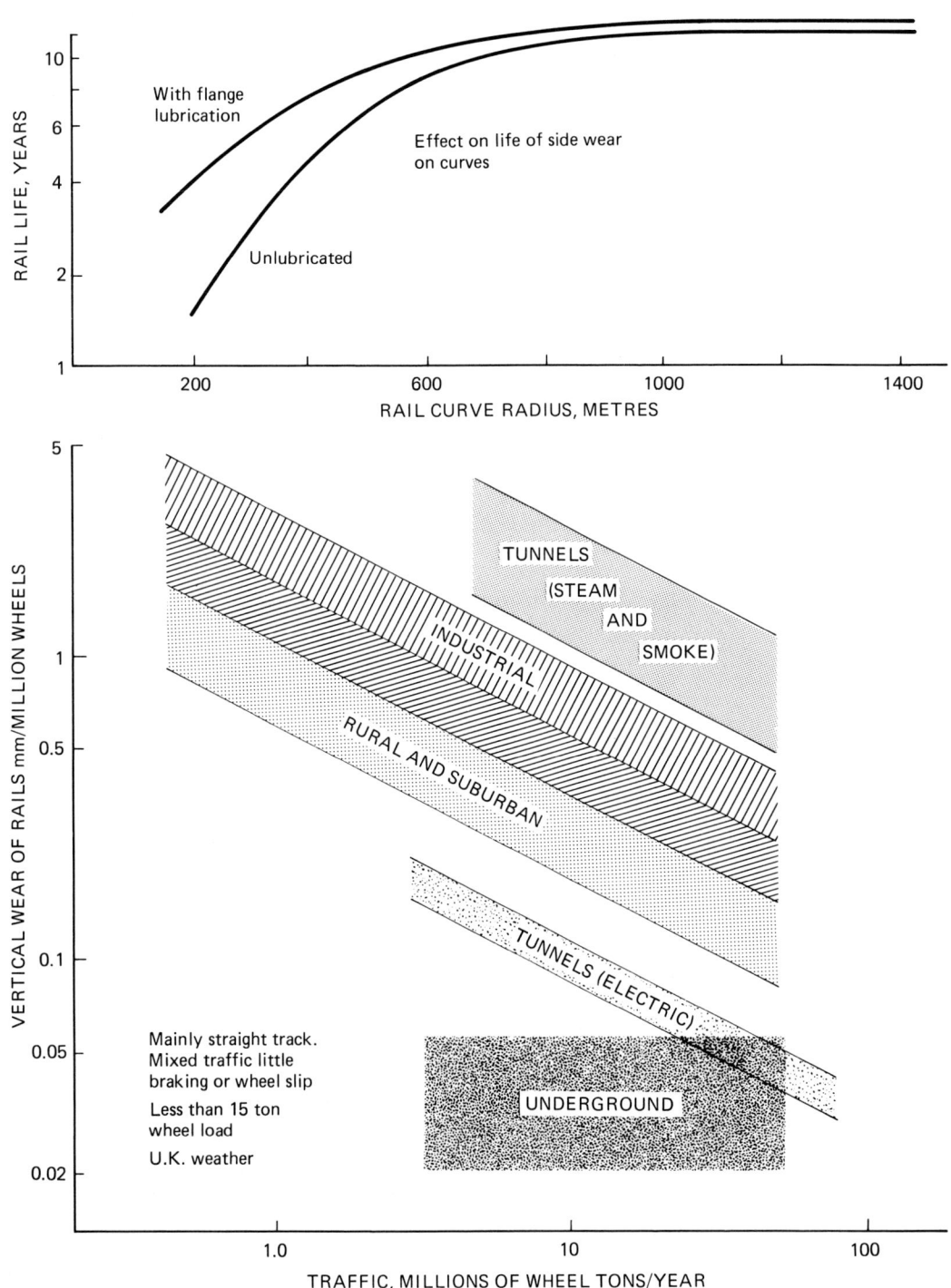

Figure 11.2 The effect of traffic density on rail wear is plotted to show the influence of the environment and type of traffic. Level of atmospheric pollution is an important controlling factor. Rail side wear is one to three times greater than vertical wear on straight track, and increases as the radius of the curve diminishes below 1000 metres

Wheels, rails and tyres — 11

Table 11.2 The effect of various factors on load capacity

	Speed (V)	Elevated temp.	Wheel dia.	Material composition	Inflation pressure	Wear	Other factors
Steel tyres on steel rails	Small effect only	Not important except in the extreme. High temperatures generated during braking can cause thermal fatigue of tyre treads	See Figure 11.4	Hardened steels or increased 'carbon equivalent' content increase load capacity	Not applicable	Wear will reduce the section modulus and hence the load capacity of the rail	
Pneumatic rubber tyres	The effect of speed on load capacity can be reduced by increasing inflation pressure, see Figure 11.5	Load capacity must be limited to prevent excessive temperature rise. Maximum temperature 120°C. Increased inflation pressure reduces running temperature	See Figure 11.1. Outside diameter and tread width can vary widely for a particular rim size	Load capacity increases with tyre ply rating. Materials with low internal friction are more suitable for consistent high-speed use	Inflation pressure is a vital factor affecting load capacity, see Figure 11.1 and other columns	Within the legal limits wear has no effect provided the tyre has been correctly used and is undamaged	

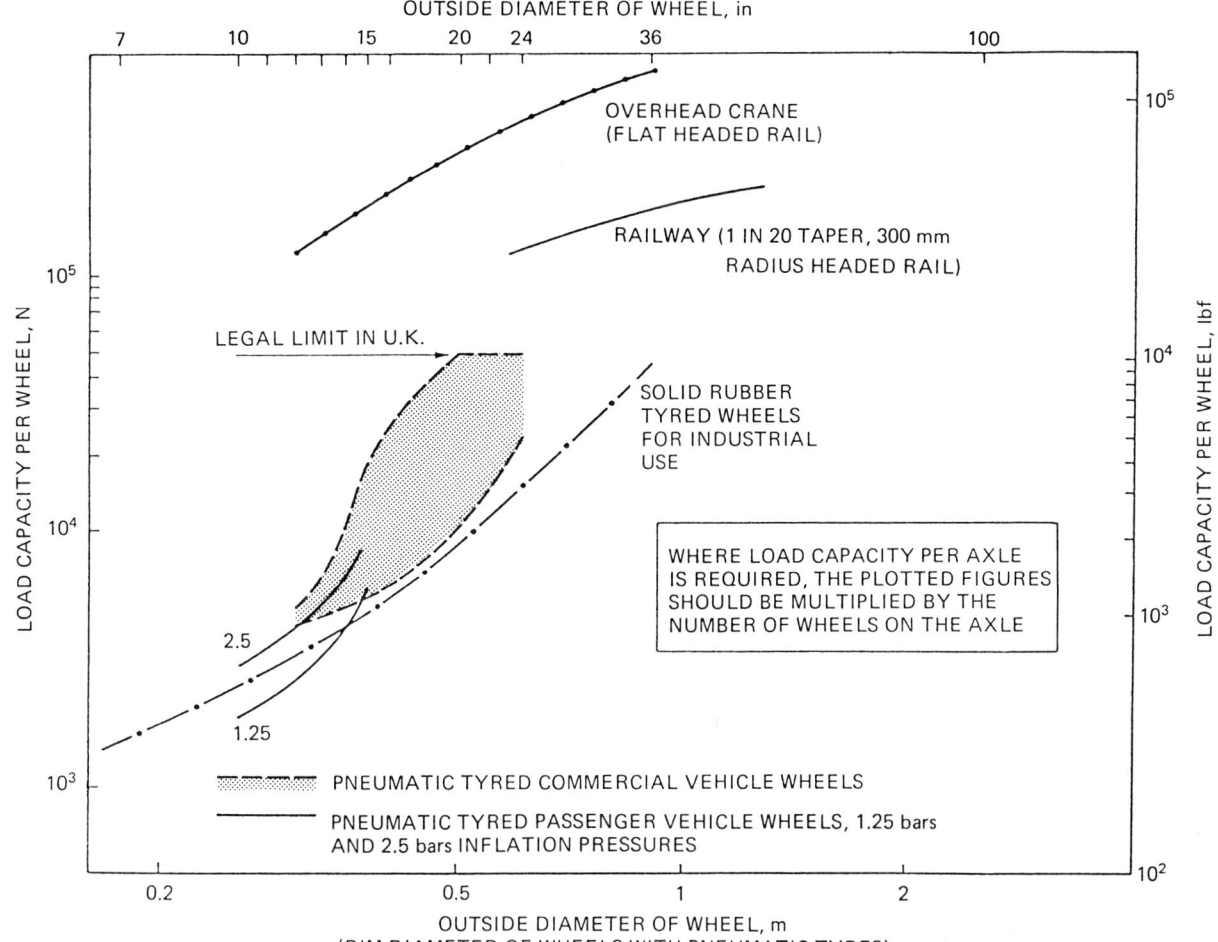

Figure 11.3 *Load capacity is plotted against wheel size for typical UK operating conditions.* Steel rail material to ISO 5003 (0.45% C 1% Mn). Steel tyre material to BS 970 Pt 2 1988. Pneumatic tyres are of cross ply construction to ISO 4251 and 4223. Solid rubber tyres are to American Tyre and Rim Association formula taking width and thickness as $\frac{1}{4}$ and $\frac{1}{16}$ outside diameter respectively

11 Wheels, rails and tyres

Table 11.2 (continued)

	Speed (V)	Elevated temp.	Wheel dia.	Material composition	Inflation pressure	Wear	Other factors
Solid tyres	Load capacity roughly inversely proportional to speed with rubber tyres	Important limiting factor. Maximum dependent on tyre material, e.g. 120°C for rubber tyres	Load capacity proportional to (diameter)$^\beta$ where $\beta = 0.5$ to 1.25	Load capacity is dependent on material, physical and mechanical properties, such as yield strength and Young's modulus	Not applicable	Tyre damage, i.e. bond failure, cracking, cutting and tearing limit load capacity	Load capacity reduced if wheel is driven

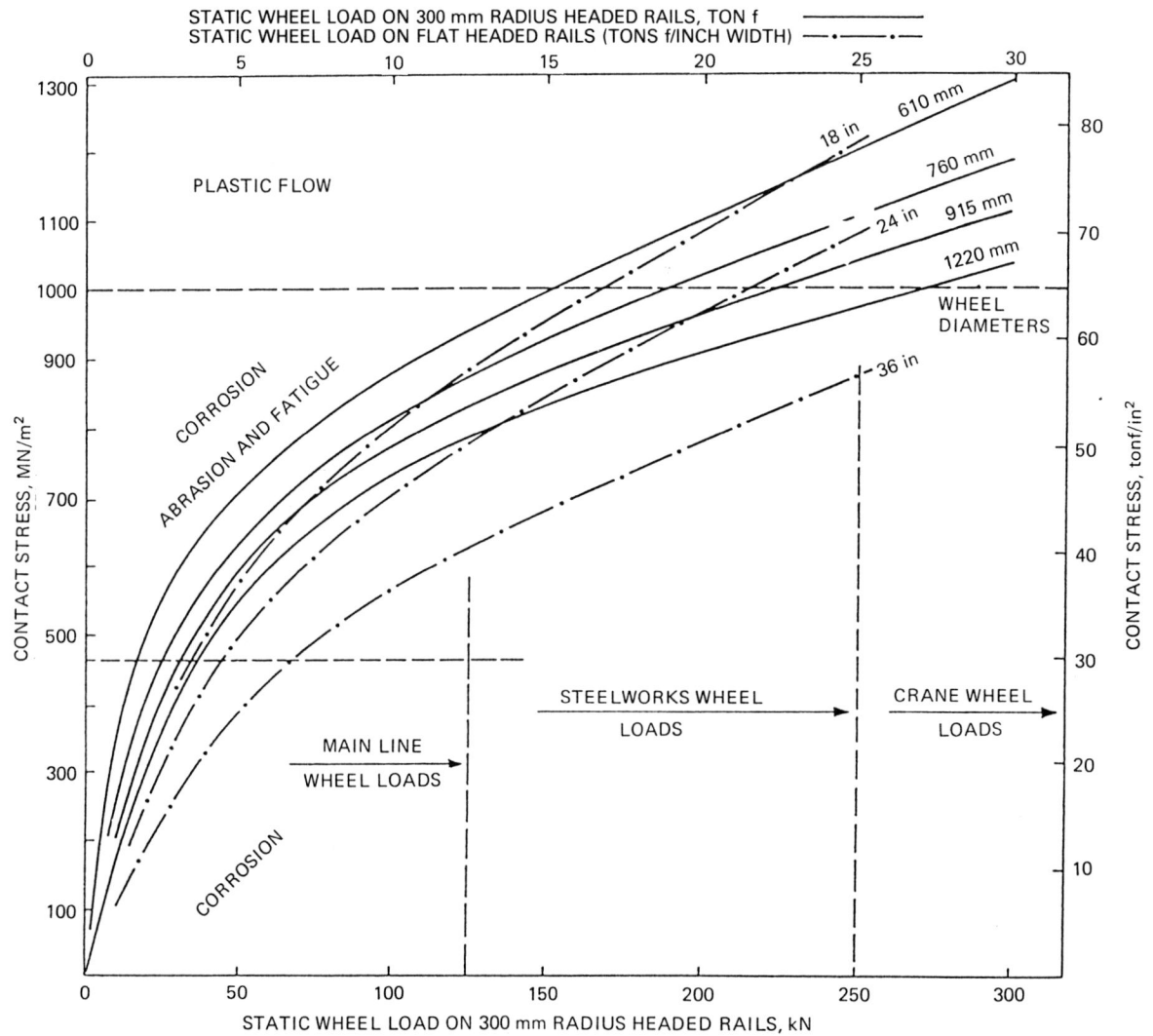

Figure 11.4 Rail contact stress and its dependence on static wheel load and wheel diameter is shown for flat-headed crane wheels and a typical main-line rail with 300 mm head radius. *The predominant wear mechanisms over the ranges of stress are shown. Wear of main-line rails typically takes the form of corrosion followed by abrasion. Fatigue cracking and plastic deformation become important where load and traffic density are high.*

Wheels, rails and tyres

Table 11.3 The effect of various factors on adhesion or traction (T), skidding (S) and rolling resistance (RR)

	Load (W)	Speed	Elevated Temp.	Water	Wheel dia. (D)	Wheel width (ω)	Material	Mating surfaces	Others
Steel tyres	$RR \alpha W^{0.9}$ $T \alpha W$ as a first approximation	Adhesion decreases slightly with speed, see Figure 11.6. Increased resistance with speed results mainly from the effect of rail joints and suspension characteristics	Little effect, except where very high local temperatures are applied to burn off adhesion-limiting surface contaminants on wheel and rail	Light rain has a marked deleterious effect on adhesion. Continuous heavy rain can improve adhesion by cleaning the rail surface	$RR \alpha D^{-\gamma}$ where $\gamma = 0.5$ to 1	Little effect	Little effect	Adhesion can be increased by removing contaminants from the surface of the rail or by sanding	Diesel- or electric-driven wheels have greater adhesion than steam because of the smoother torque
Pneumatic rubber tyres	RR increases with load unless inflation pressure is also increased. $T \alpha W$ as a first approximation	See Figure 11.6. Adhesion reduced by the effect of sideways forces α (speed)^2 in cornering	Rubber friction coefficient decreases with temperature – about 10% per 15°C change	Large reduction in friction, see Figure 11.6. Some rubbers are less affected than others	Adhesion increases, while rolling resistance decreases with increase in diameter	Rolling resistance decreases slightly with width with modern tyre designs. Traction increases with width	Radial ply construction gives improvements in all three coefficients. Rubber composition is more important than tread pattern	On wet roads harshness is more important than roughness except at high speeds, see Figure 11.6. Road surface is a more important factor than tyre condition	Aquaplaning speed α (inflation pressure)$^{1/2}$. Wear down to 2 mm tread depth has little effect on adhesion: smooth treads give as little as half the adhesion of unworn patterned treads on poor, wet surfaces
Solid tyres	$RR \alpha W^{4/3}$ $T \alpha W$ as a first approximation	Not suitable for moderate high speeds (above 40 km/h) unless specially designed	Dependent on properties of tyre material	Affected in a similar manner to pneumatic tyres, but more so	$RR \alpha D^{-0.6}$	$RR \alpha \omega^{-0.3}$	$RR \alpha E^{1/2}$ (E is Young's modulus)	Solid tyres only suitable for smooth surfaces	

11 Wheels, rails and tyres

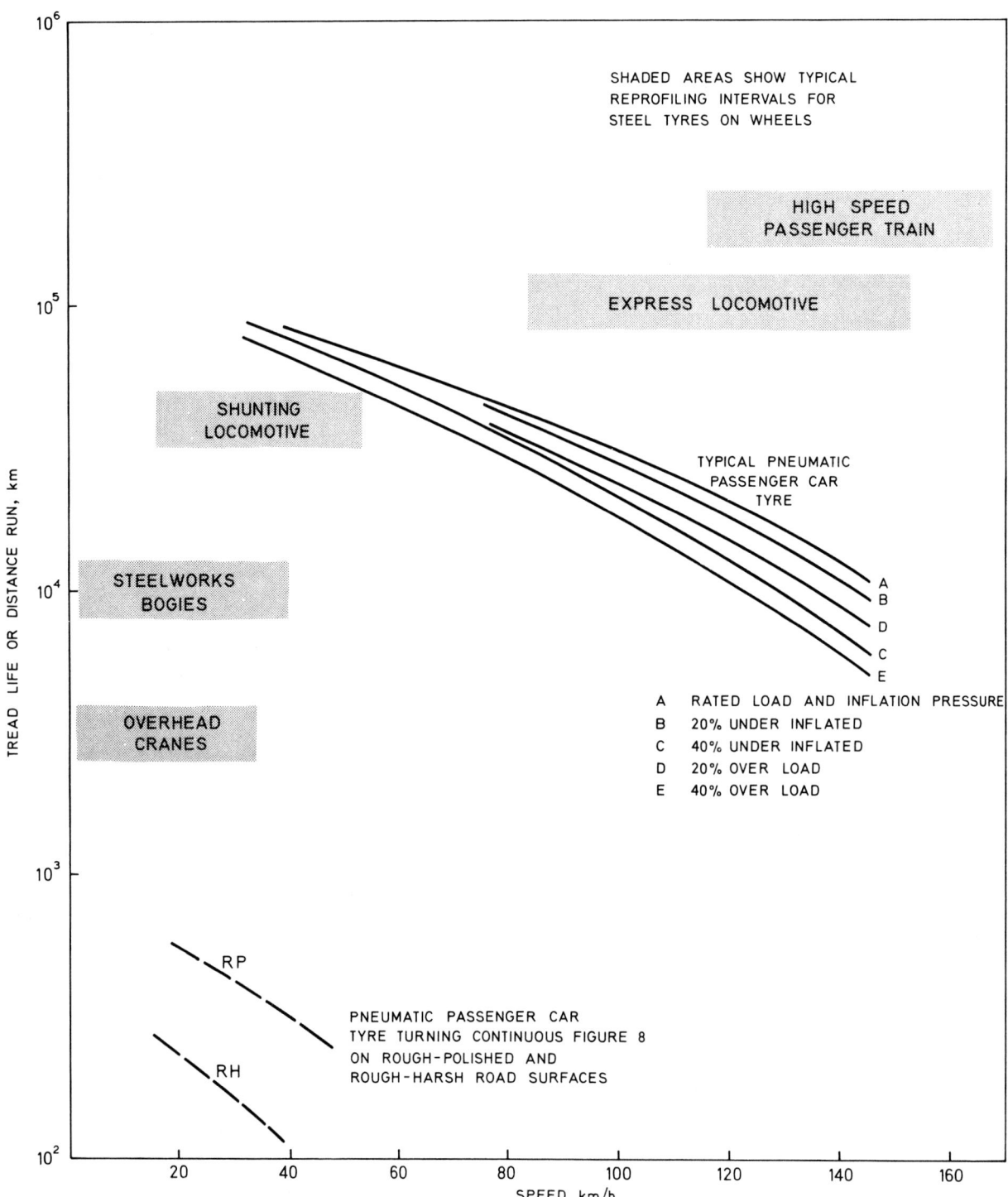

Figure 11.5 *Wheel life or distance run between reprofiling intervals is plotted against speed for overhead cranes, different types of railway rolling stock, and passenger-car pneumatic tyres*

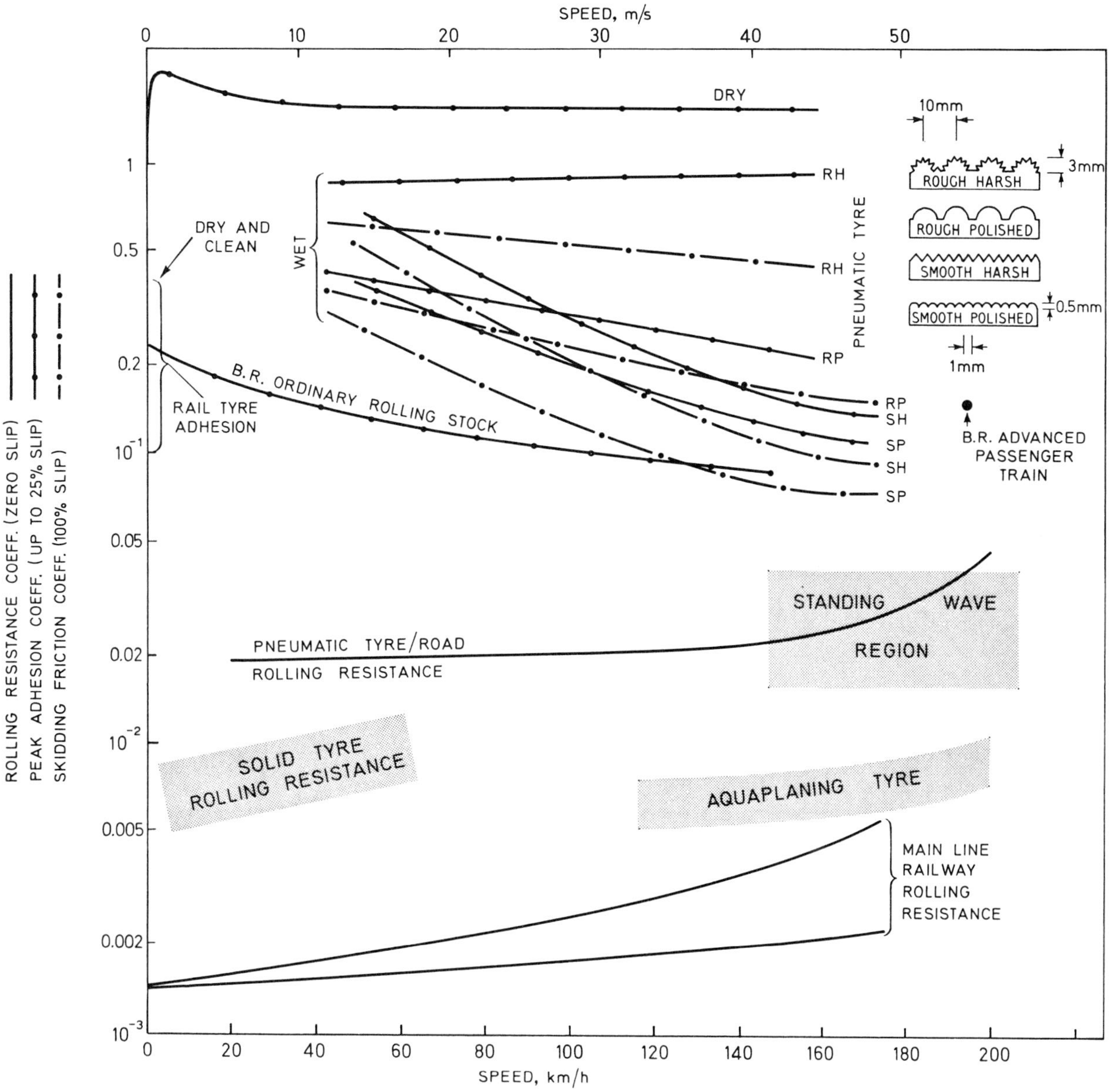

Wheels, rails and tyres 11

Figure 11.6 Rolling resistance, peak adhesion or traction, and skidding friction coefficients are plotted against speed for passenger-car tyres on wet and dry road surfaces of various typical characteristics, for solid tyres and for main-line railway rolling stock

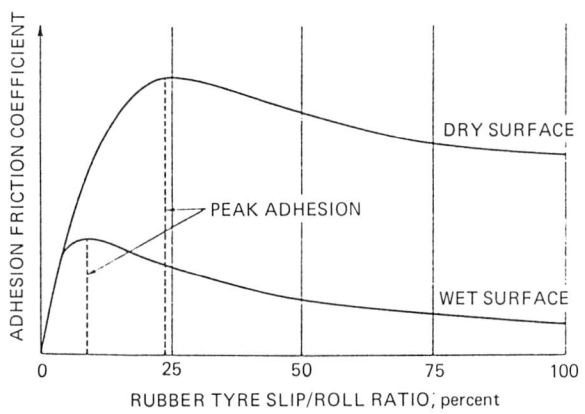

Figure 11.7 Typical adhesion versus tyre slip/roll ratio for treaded rubber tyres on wet and dry surfaces. 0 per cent signifies pure rolling, 100 per cent pure sliding

12 Capstans and drums

Capstans and drums are employed for rope drives. The former are generally friction drives whilst the latter are usually direct drives with the rope attached to the drum. The roles, however, may be reversed. Friction drives control the rope motion by developing traction between the driving sheave and the rope and might be preferred to direct drives for reasons of economy (smaller drive sheave), safety (slippage possible) or necessity (e.g. endless haulage).

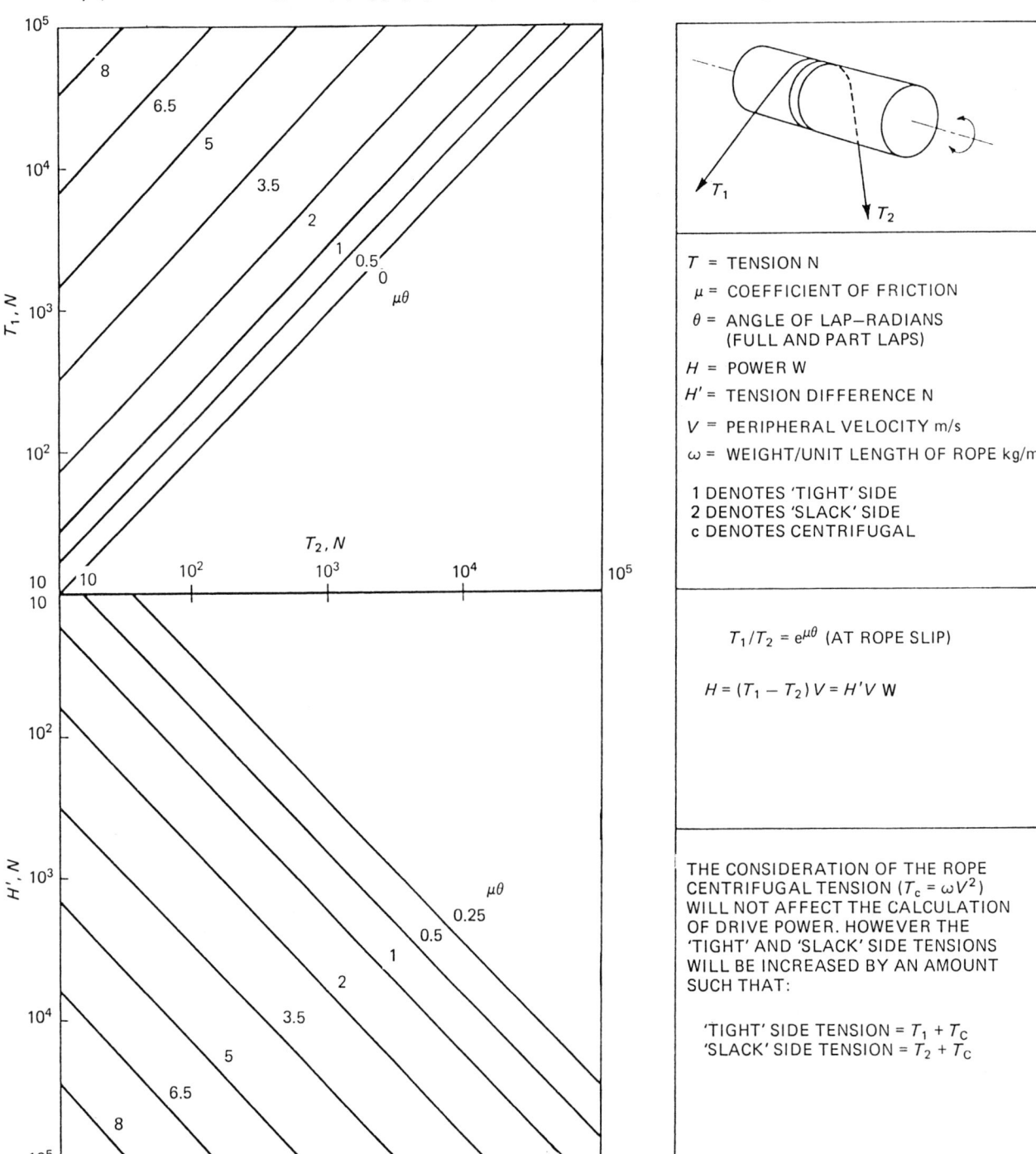

T = TENSION N
μ = COEFFICIENT OF FRICTION
θ = ANGLE OF LAP—RADIANS (FULL AND PART LAPS)
H = POWER W
H' = TENSION DIFFERENCE N
V = PERIPHERAL VELOCITY m/s
ω = WEIGHT/UNIT LENGTH OF ROPE kg/m

1 DENOTES 'TIGHT' SIDE
2 DENOTES 'SLACK' SIDE
c DENOTES CENTRIFUGAL

$T_1/T_2 = e^{\mu\theta}$ (AT ROPE SLIP)

$H = (T_1 - T_2)V = H'V$ W

THE CONSIDERATION OF THE ROPE CENTRIFUGAL TENSION ($T_c = \omega V^2$) WILL NOT AFFECT THE CALCULATION OF DRIVE POWER. HOWEVER THE 'TIGHT' AND 'SLACK' SIDE TENSIONS WILL BE INCREASED BY AN AMOUNT SUCH THAT:

'TIGHT' SIDE TENSION = $T_1 + T_c$
'SLACK' SIDE TENSION = $T_2 + T_c$

FRICTION DRIVES

The figure shows the determination of the approximate performance of friction drives at the rope slip condition. In contrast to the belt and pulley situation the required 'tight' or 'slack' side tension is usually known. Capstans are widely used with vegetable, animal or man-made fibre ropes, but more rigorous conditions demand wire rope.

Capstans and drums

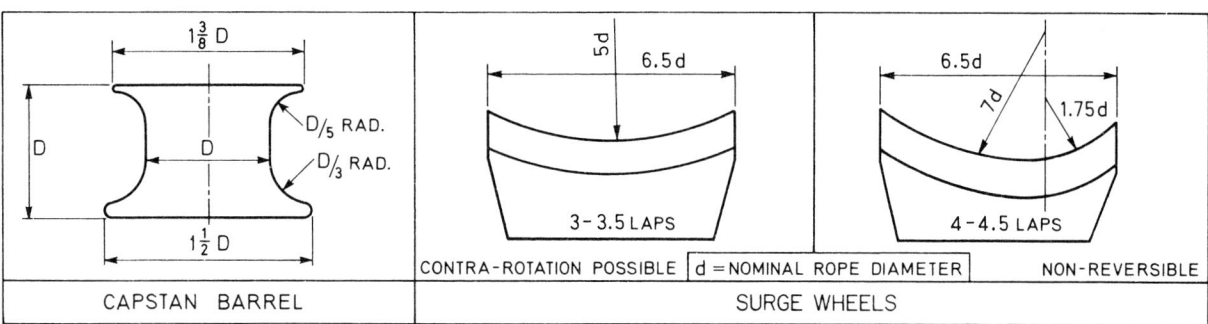

CAPSTAN BARREL	SURGE WHEELS
	CONTRA-ROTATION POSSIBLE — d = NOMINAL ROPE DIAMETER — NON-REVERSIBLE

The figure shows the typical profiles of a capstan barrel and surge wheels. Grooving is not appropriate. Capstans are often employed when relatively low rope tensions are involved and hence must have a large flare, which whilst ensuring free movement does not allow disengagement of the rope. Surge wheels are used on endless haulage systems with wire ropes. The large rope tensions involved mean that only a moderate flare is necessary. The laps slip or surge sideways across the surface as the rope moves on and off the wheel, hence the term surge wheel. This movement necessitates differing wheel shapes depending upon the rotational requirements.

GROOVE PROFILE	PLAIN	U	V (α TYPICALLY 40°)	UNDERCUT U
APPARENT COEFFICIENT OF FRICTION μ'	μ	$4\mu/\pi$	$\mu / \sin \frac{\alpha}{2}$	$\dfrac{4\mu(1 - \sin \frac{\alpha}{2})}{\pi - \alpha - \sin \alpha}$

The figure shows grooving in friction drives to increase tractive effort. The apparent coefficient of friction, μ', replaces the normal coefficient in calculations.

Rope material	Drive sheave or sheave liner material	Friction coefficient at slip (dry conditions)
Wire	Iron or steel	0.12
	Wood	0.24
	Rubber or leather	0.50
Nylon	Aluminium	0.28
	Iron or steel	0.25
Woven cotton	Iron or steel	0.22
Leather	Iron or steel	0.50

For wet conditions reduce the friction coefficient by 25%, for greasy conditions 50%

The table gives an approximate guide to the value of friction coefficient at slip for various rope and driver material combinations. A factor of safety reducing the values shown and appropriate to the application is usually incorporated.

DIRECT DRIVES

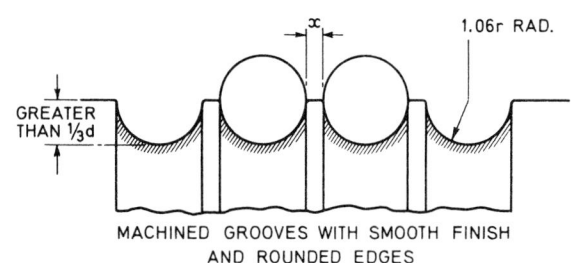

MACHINED GROOVES WITH SMOOTH FINISH AND ROUNDED EDGES

ROPE NOMINAL DIAMETER d	CLEARANCE BETWEEN TURNS x
UP TO 13 mm	1.6 mm
UP TO 28 mm	2.4 mm
UP TO 38 mm	3.2 mm

$d = 2r$ = NOMINAL ROPE DIAMETER
x = CLEARANCE BETWEEN TURNS

The figure shows typical grooving of a wire rope drum drive with the rope attached to the drum. Performance is unaffected by frictional considerations. Pinching of the rope is avoided where grooves are employed for guidance purposes. Drum grooves are normally of cast iron, carbon steel or alloy steel and reduce wear of both drum and rope.

13 Wire ropes

SELECTION OF LOAD CARRYING WIRE ROPES

Table 13.1 Ropes for industrial applications

Application	Conveyors, small hoists, small stays, trawl warps*	Excavators (drag), lifts,* sidelines (dredging), mast stays, pendants, crane ropes (small)	Lifts,* drilling lines, scrapers, trip ropes, skip hoists, pile driving	Crane hoists (multi-fall), grabs, slings	Tower cranes (hoist), mobile cranes (small), drop ball cranes, boatfalls	Deck cranes (single fall), mobile cranes (large)
Construction	6 × 7(6/1) IWRC	6 × 19(9/9/1) IWRC	6 × 19 (12/6 + 6F/1) IWRC	6 × 36 (14/7 and 7/7/1) IWRC	17 × 7(6/1) FC	12 × 6 over 3 × 24
MBL kN	$d^2 \times 0.633$	$d^2 \times 0.632$	$d^2 \times 0.643$	$d^2 \times 0.630$	$d^2 \times 0.562$	$d^2 \times 0.530$
Wt kg/100 m	$d^2 \times 0.382$	$d^2 \times 0.398$	$d^2 \times 0.408$	$d^2 \times 0.406$	$d^2 \times 0.372$	$d^2 \times 0.362$
E N/mm^2	68 600	64 700	64 700	64 700	53 900	53 900

d = nominal diameter of rope
FC = Fibre Core
E = Modulus of Elasticity of the rope (differs from elastic modulus of wires in rope due to helical formation of latter)
MBL = Minimum Breaking Load
IWRC = Independent Wire Rope Core

Notes: Minimum Breaking Loads based on a tensile strength of 1.8 kN/mm^2

With Fibre Cores: Reduce *MBL* by 8% approximately.
Reduce weight by 9% approximately.

Construction and fatigue performance

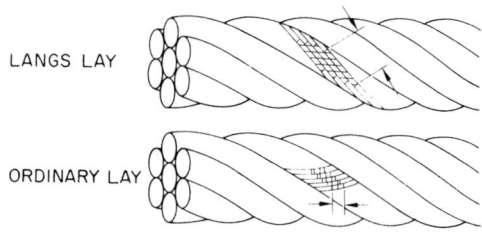

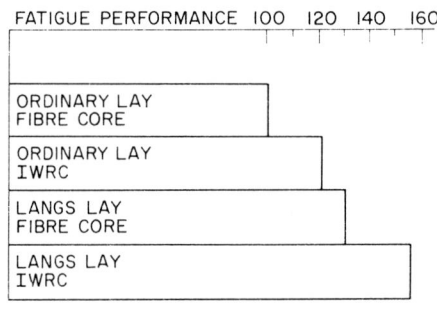

Figure 13.1 Comparison of ropes in fatigue based on a factor of safety of six

Wires in a Langs Lay rope are laid in the same direction as the strands, thus rope in this construction can permit a greater reduction in volume of steel from the rope surface for the same loss in breaking load than is possible with an ordinary lay rope.

Langs Lay ropes have greater resistance to abrasion, and are also superior to ordinary lay ropes in fatigue resistance. However, they develop high torque, and cannot be used when this factor is of importance.

Wire ropes 13

Table 13.2 Ropes for mining applications

Application	Haulage (general), conveyors, winding (small sizes), cableway haulage	Winding	Winding—drum hoists, friction hoists	Shaft guides	Balance for drum and friction hoists	Aerial carrying ropes, cableways
Construction	6 × 8(7/1)Δ flattened strand 160 Grade	6 × 25 12/12/1 Br 9/3 flattened Strand 180 Grade	Locked coil—winding	Locked coil—guide	'Superflex' 20 × 6/17 × 6/ 13 × 6/6 × 19 110 Grade	Locked coil—aerial
MBL kN	$d^2 \times 0.565$	$d^2 \times 0.613$	$d^2 \times 0.851$	$d^2 \times 0.500$	$d^2 \times 0.368$	$d^2 \times 0.809$
Wt kg/100 m	$d^2 \times 0.407$	$d^2 \times 0.413$	$d^2 \times 0.563$	$d^2 \times 0.550$	$d^2 \times 0.385$	$d^2 \times 0.568$
E N/mm²	61 800	58 800	98 100	117 700	53 900	117 700

Loading and performances

Static load $(T_S) = W_C + W_L + W_R$

Static factor of safety = $MBL : T_S$ (i.e. $FOS = 6.5:1$)
for cranes etc. W_R is not considered

Dynamic load $(T_D) = T_S \cdot a$

Bending load $(T_B) = \dfrac{E \Delta A}{D}$

Total load $(T_{max}) = T_S + T_D + T_B$

Percentage stress $= \dfrac{T_{max} \times 100}{MBL}$

(See Figure 13.2 for stress reversals based on percentage stress.)

Elastic stretch $= \dfrac{T_S \cdot L}{EA}$

W_C = Weight of conveyance and attachments
W_L = Payload
W_R = Weight of suspended rope
a = Acceleration
g = Gravitational constant
E = Elastic Modulus (based on rope area)
Δ = Diameter of rope at centre of outer wires
A = Area of rope
D = Diameter of drum or sheave
L = Rope length

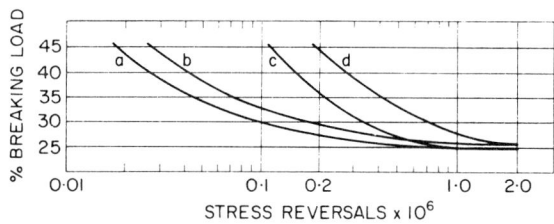

Figure 13.2 Fatigue tests on wire ropes (a) flattened strand, winding (b) locked coil, winding (c) 6 × 25 RS IWRC ordinary lay (d) 6 × 36 IWRC ordinary lay

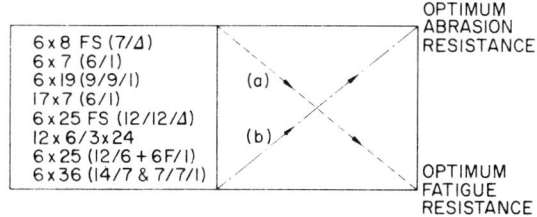

Figure 13.3 Comparison of construction for (a) resistance to bending fatigue (b) resistance to abrasion and crushing

14 Control cables

SELECTION OF CONTROL CABLES AND WIRES

The form of cables and wires which can be used for control purposes can be defined as follows:

Wire—a single 'solid' circular section i.e. piano or music wire.

Single strand—an assembly of wires of appropriate shape and dimensions, spun helically in one or more layers around a core.

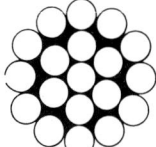

19 WIRES

Multi strand—an assembly of strands spun helically in one or more layers around a core.

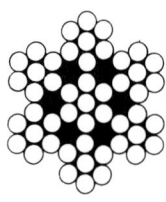

49 WIRES

Wire can be used in push and pull conditions inside suitable conduits (outer casings). Single strand cable should be used inside conduits for the transmission of tension loads. Rope constructions are used around pulleys. Single strand cable reinforced with a helically close wound steel tape around the periphery of the strand provides sufficient resistance to buckling for use within suitable conduits to transmit both push and pull loads.

Load capacity

Minimum breaking loads of cable taking into account the helical effects of their construction are approximately 85% of the calculated aggregate breaking load. Considerations of safety factors to cater for possible intermittent increased load conditions modify considerably the load capacity of cables from their minimum breaking load values to practical load values for good design. The table gives a recommended typical selection of single strands in relation to their use within conduit.

Table 14.1 Performance of stranded cables

Strand construction	Overall diameter mm	Minimum breaking load N	Conduit type	Overall diameter conduit mm	Conduit minimum bend radius mm	Recommended maximum input load N
19 WIRES	1.53	2220	Single wire section	4.45	54	220
	2.00	3340	Twin wire section	8.10	127	510
	2.80	8000	Twin wire section	9.80	150	1780
	3.50	12 900	Twin wire section	11.40	170	3380
36 WIRES	5.50	28 900	Twin wire section	16.40	250	8400

Notes: Overall diameter of strand excludes thermoplastic covering. Conduits incorporate appropriate thermoplastic liners. Minimum bend radii for conduits may well vary from the above values when the conduit construction comprises long lay wires.

Control cables

Pulley size and cable fatigue

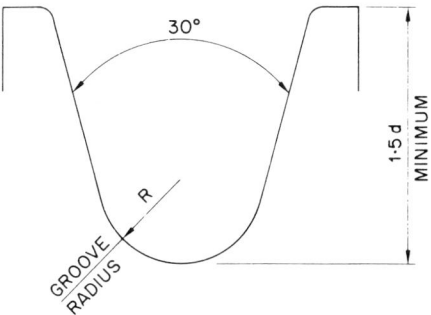

Figure 14.1 Pulley groove form for control purposes. Groove radius R = 0.53d where d = cable diameter inclusive of any thermoplastic covering

Although Bowden cables were conceived to eliminate the use of ropes and pulleys, control configurations using steel wire ropes and pulleys are in use.

For very small rope below 2 mm diameter, pulley diameter (over groove) should not be less than thirty times rope diameter. For small rope used as controls (2 to 6 mm diameter), pulley diameter over groove should not be less than thirty-five times actual rope diameter (exclude the effect of any plastic covering). Nylon or acetal moulded pulleys give excellent life, resist corrosion and reduce friction and wear between rope and pulley. Prediction of fatigue life of ropes around pulleys is difficult and involves many factors. Practical tests have shown that provided the minimum recommended pulley diameters are used and that the tension load in the rope does not exceed 10% of the minimum breaking load of the rope, long life can be expected. Angles of wrap between 10° and 30° should be avoided; reverse bends should be avoided wherever possible.

The following table shows the effect on fatigue life with varying

$$\frac{\text{pulley dia. over groove } (D)}{\text{rope dia. } (d)}$$

on a single pulley. 7 × 7 rope 2.60 mm (d), nylon covered to 3.60 mm. Pulley—nylon; axle mild steel, grease lubricated. 90° angle of wrap. Tension load in rope 625 N. Minimum breaking load of rope 6250 N.

Table 14.2 Effect of diameter ratio on fatigue life

D/d	Cycles to failure
20	56 000
25	170 000
30	330 000
37.5	after 650 000 no failure

Friction in Bowden cables

The friction of wires and cables inside conduits (Bowden cables) depends on the materials and construction of the cable and conduit and on the bend radii obtained when the system is installed. Control runs employing flexible conduit should take a natural path between the conduit grounding points.

The materials used may be:

Metal cable in a metal conduit with grease lubrication.
Metal cable in a plastic lined conduit.
Plastic coated metal cable in a conduit lined with a dissimilar plastic.

The last combination gives low friction and high efficiency compared with the first, as shown in the figure below.

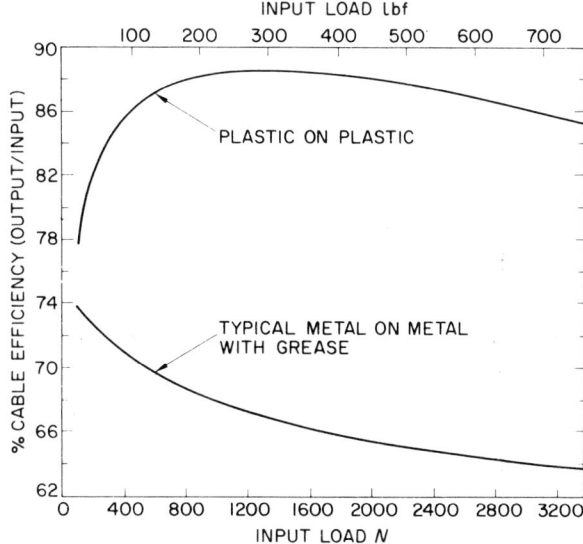

Figure 14.2 Variation of cable efficiency with input load. Configuration: 180° 'U' bend 210 mm radius and conduit length 800 mm. For plastic on plastic cables with 90° bend add 2% to the above values of efficiency. With 270° bends deduct 5% from the above values of efficiency

15 Damping devices

GENERAL CHARACTERISTICS

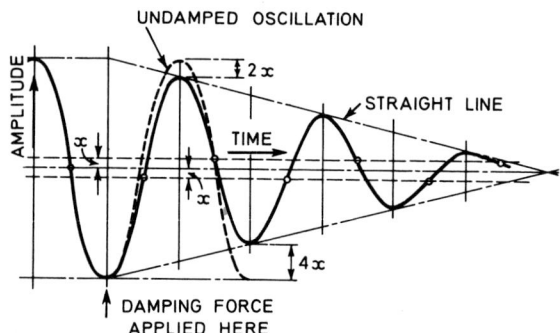

Figure 15.1 The decay of oscillations by friction damping

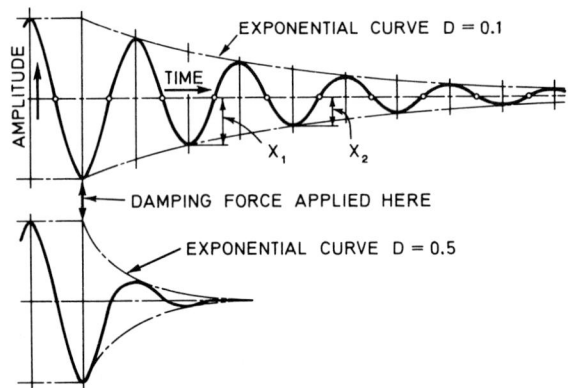

Figure 15.2 The decay of oscillations by viscous hydraulic damping

Damping devices are used to provide forces to resist relative motion and oscillation. The two main types are friction dampers, using friction between solid components and hydraulic dampers using mainly viscous effects.

If an oscillation is damped with a friction damper, the oscillation decay will follow a straight line (Figure 15.1). The amplitude of each successive oscillation will be reduced by $4x$
where $x = F/k$

and F = damping force (newtons)

k = system spring rate (newtons/metre)

With a hydraulic damper, the decay of the oscillation follows an exponential curve (Figure 15.2).

The diagram of the damping action of a hydraulic damper shows the effect of different damping factors. These are related to the number of cycles that it takes for the system to come to rest after an impact displacement. This relationship can be used to make an experimental measurement of the damping factor of an existing system.

The relationship between damping force and velocity varies with the type of damper and can conveniently be described by the formula

$$F = cV^n$$

where c = a constant which depends on factors such as the size of the damper

V = the displacement velocity

n = a constant which depends on the working principle of the damper

Type of damper	Value of n	Force characteristic
Friction damper	0	F is constant for all values of V
Hydraulic damper with constant area flow passages	2	F is proportional to V^2
Hydraulic damper with valves to control the flow	1	F is proportional to V

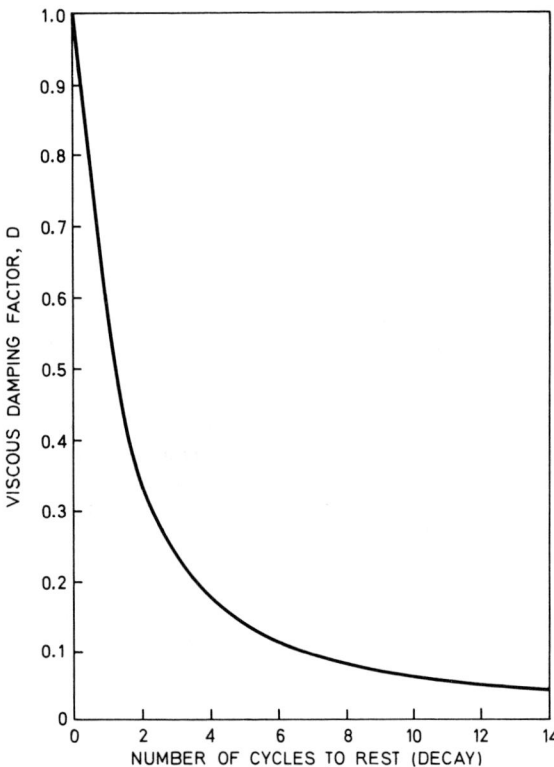

Figure 15.3 The effect of viscous hydraulic damping factor on the number of cycles required to come to rest after a single impact

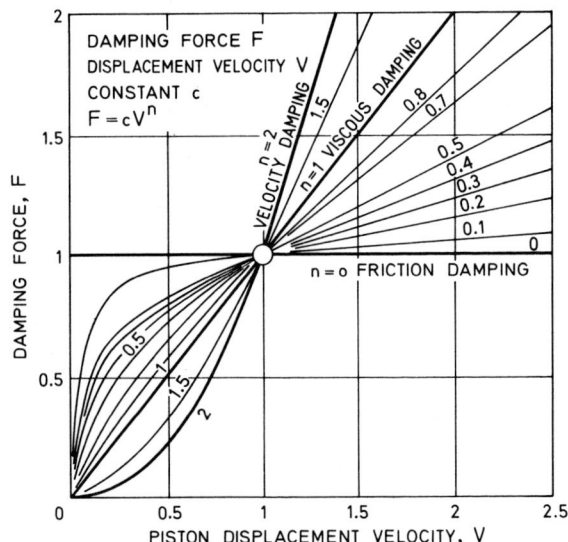

Figure 15.4 The dependence of the damping force F on the piston velocity V at various values of the exponent n

Damping devices 15

PERFORMANCE OF HYDRAULIC DAMPERS FITTED WITH VALVES

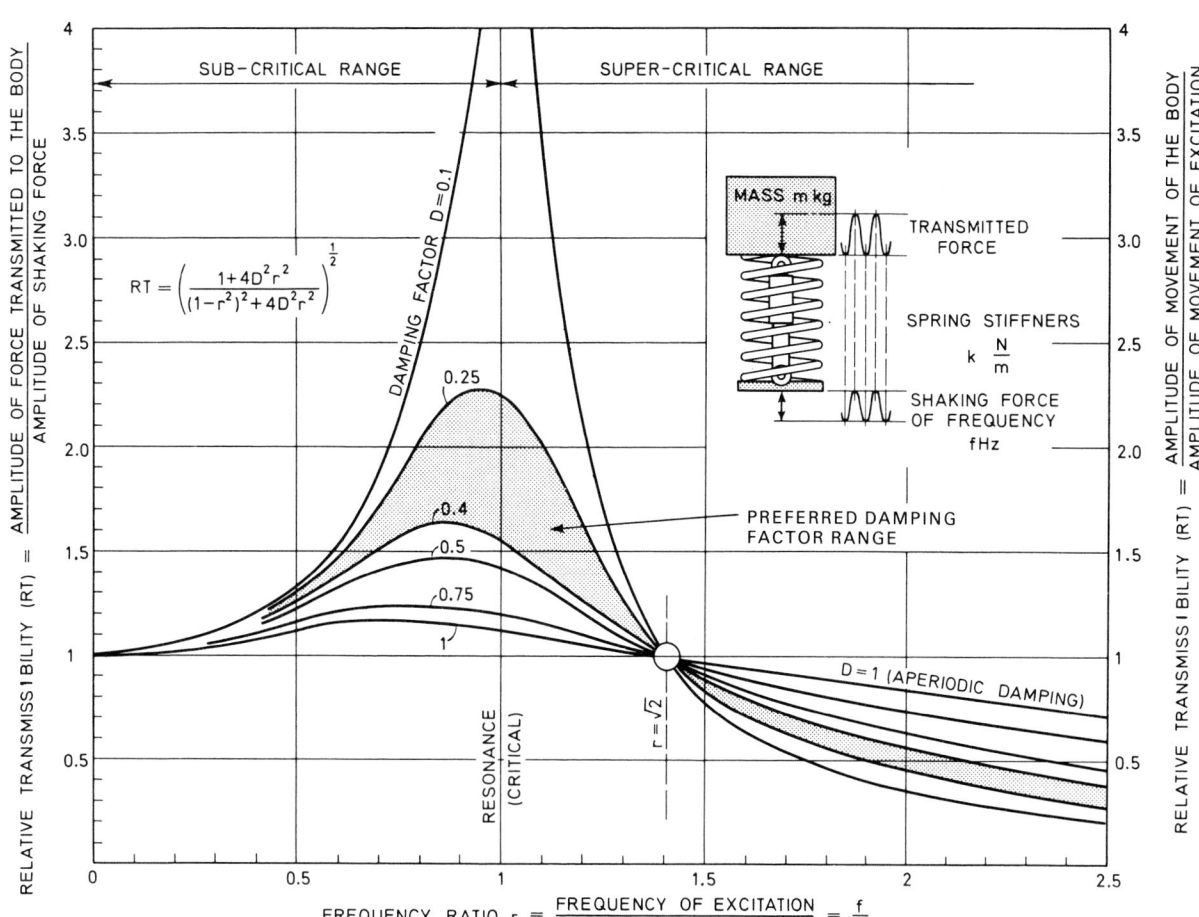

Figure 15.5 Curves of relative transmissibility for a number of representative viscous damping factors

If two components are connected by a spring system, such as a wheel and the body of a vehicle, the forces and oscillating movement transmitted by one to the other, will depend on the frequency of the oscillations being transmitted in relation to the natural frequency of the system, and the damping factor of any damping device which is incorporated.

In a viscous hydraulic damper, in which F is proportional to V, the damping factor D is given by

$$D = \frac{C_1}{4\pi \, m f_r}$$

where C^1 = damping force per unit velocity N/ms^{-1}
m = mass of the critical component kg
f_r = natural (resonant) frequency of the system
$\frac{1}{2\pi}\left(\frac{k}{m}\right)^{1/2}$ Hz

This damping factor is not affected by velocity for a damper of this kind, and consequently the relative transmissibility of forces and movements will depend mainly on the frequency ratio r, of the excitation frequency to the resonant frequency.

For a system of this kind with resonance at $r = 1$, damping is desirable over the range $r = 0$ to $r = 1.4$. If r is normally greater than 1.4 low D values become more desirable. For conditions extending over a wide range of r, values of $D = 0.25$ to 0.4 are found to give the best results.

Viscous hydraulic dampers of good design are available with adjustable valves to give controlled relationships between damping force and damper velocity.

15 Damping devices

POINTS TO NOTE IN DAMPER SELECTION AND DESIGN

1. Friction dampers can potentially absorb the most energy, but are unsuitable for systems subject to steady oscillations at or near resonance, unless the friction dampers are large enough to provide a damping force of at least four times any excitation force.
2. To give satisfactory performance with acceptable wear rates, the contact pressures in friction pad dampers should not exceed 250 kN/m^2 (34 p.s.i.).
3. With hydraulic dampers, damping factors in the range 0.25 to 0.4 are generally suitable, but at frequency ratios above $r = 2.5$ the transmitted forces are higher than with friction dampers.
4. With viscous hydraulic dampers giving straight line force/velocity characteristics it is important to specify the velocity at which the force should be levelled to a constant value by a cut-off valve. A valve of adequate capacity is needed to prevent instantaneous velocity peaks from bursting the housing or causing damage to the working valves or the end mountings. The maximum allowable pressure at cut-off is usually of the order of 3.5 MN/m^2 (500 p.s.i.).
5. If damping devices are likely to be operated near the resonant frequency of the system, the rigidity of the rubber pads or mounting bushes need to be checked for adequate stiffness.

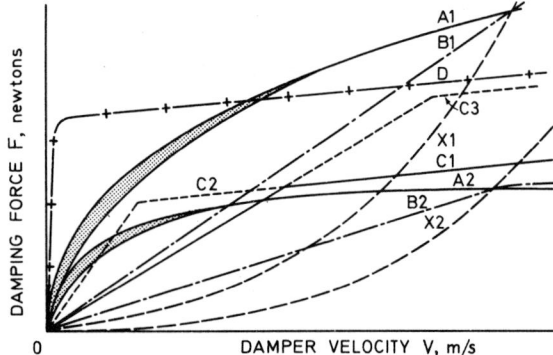

Figure 15.6 The range of force-velocity characteristics attainable with a damper of the type illustrated. These characteristics are obtained as follows:
A1 with one adjustable valve and a small diameter passage
A2 as above, but a larger diameter passage
B1 as A1 but a smaller orifice in the adjustable valve
B2 as A2 but a larger diameter orifice in the adjustable valve
C1 with two adjustable valves, the first valve controlling up to the cut-off point along lines of B1 and B2, and the characteristic above cut-off being determined by the second adjustable valve. The bore in the seat of the second adjustable valve is larger than that of the first
C2 as above, but with a different bore and orifice in the first valve
C3 as C1 but with the second valve set more closely
D with two or more adjustable valves, all set to identical values

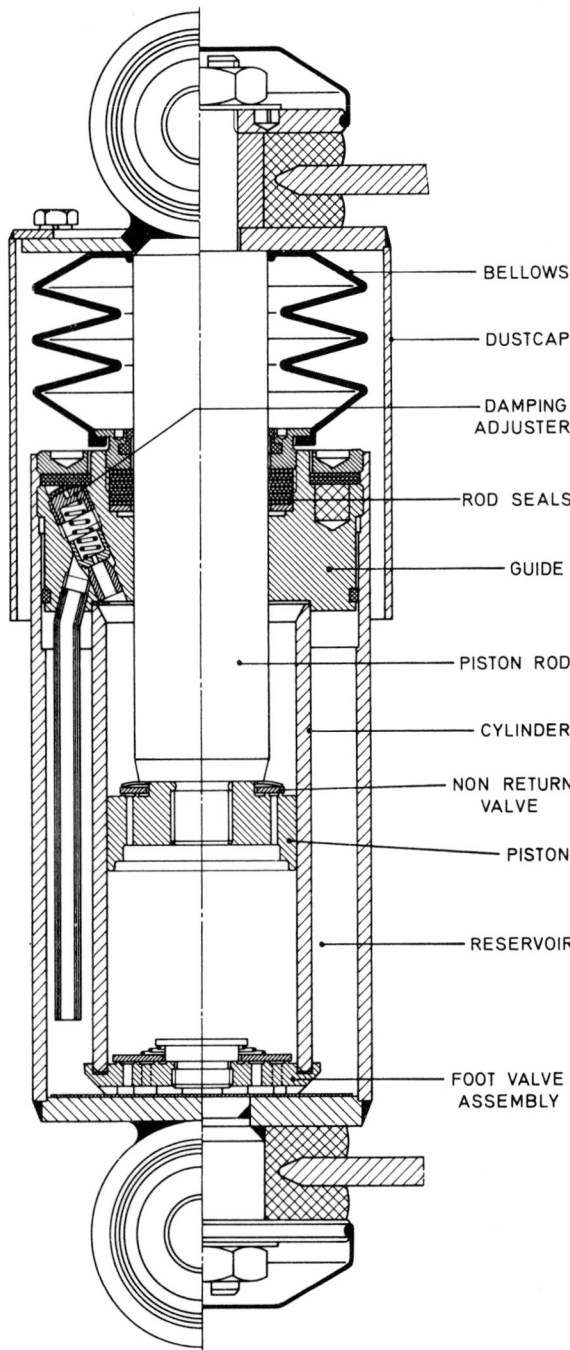

Figure 15.7 A viscous hydraulic damper. The adjustable valve provides a simple method of changing the slope of the force-velocity characteristic. The damper is intended to deal with vertical oscillation and shows alternative types of end fixing

Pistons 16

HYDRAULIC PISTONS

Type A—Semi-static operation (pressure unidirectional)

Piston may be made from aluminium or cast iron. Operation may be by a push rod or a piston rod bolted through the piston. Sealing is usually by a cup washer fitted against a flat end face.

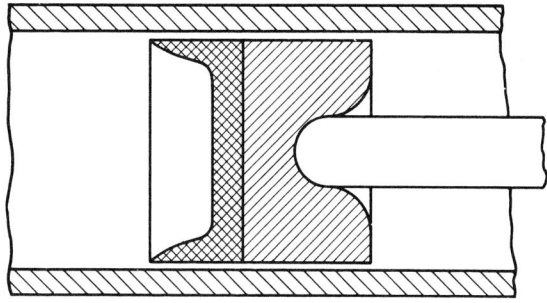

Type B—Double-acting hydraulic actuators

Pistons of aluminium, cast iron or steel, may be fitted with lip seals ('U' packings) for operating pressures up to 21 MN/m^2 (3000 p.s.i.), or PTFE or metallic rings.

Hydraulic piston with 'U' packings

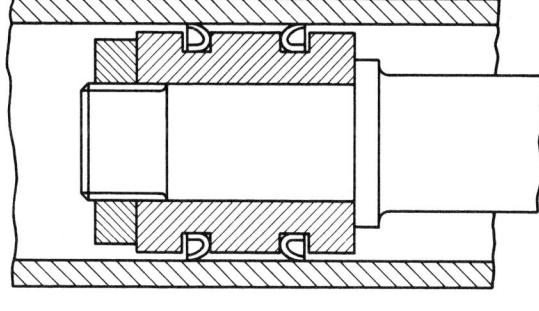

Hydraulic piston with PTFE sealing rings and bearer bands

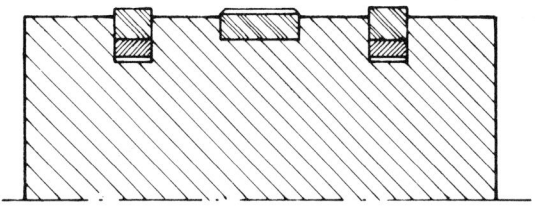

Scarf-step sealing ring arrangement, hydraulic or pneumatic: (a) unidirectional; (b) double-acting

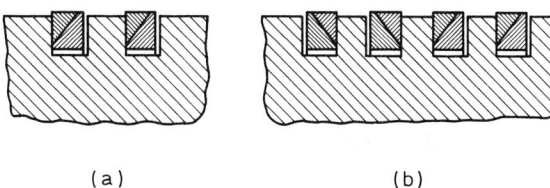

(a)   (b)

Figure 16.1 Various types of hydraulic pistons

16 Pistons

CRANK OPERATED HYDRAULIC PUMP PISTONS

Piston may be made from aluminium or cast iron, with sealing usually by metallic piston rings. The minimum strength of the piston parts should be as for IC engines.

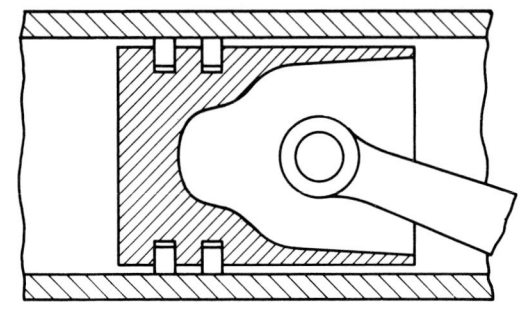

Some simple robust low-speed designs use a plain, solid cylindrical piston operating through a gland bush.

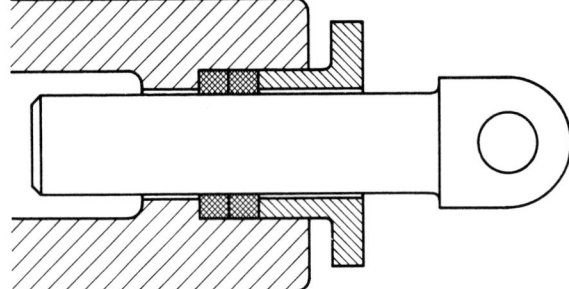

Figure 16.2 Types of pistons for hydraulic pumps

AIR AND GAS COMPRESSOR PISTONS

Piston materials may be aluminium or cast iron. Some common arrangements are shown below. Minimum strength of piston parts to be as for IC engines.

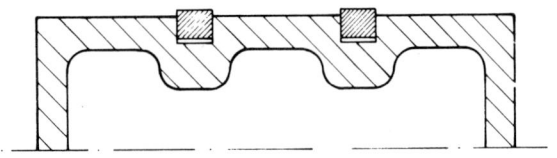

Double-acting compressor piston (lubricated) with plain piston rings

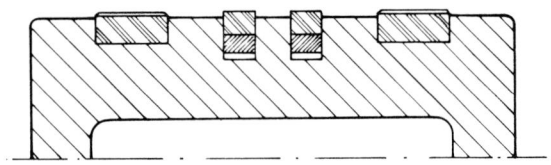

Double-acting compressor piston (non-lubricated) with PTFE sealing rings and bearer bands

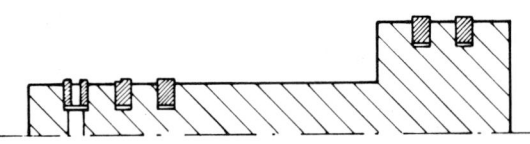

Two-stage compressor piston (lubricated)

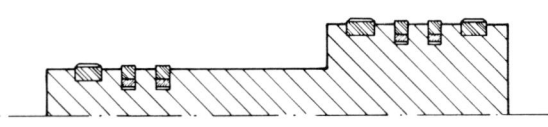

Two-stage compressor piston (non-lubricated) with PTFE sealing rings and bearer bands

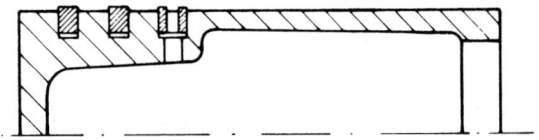

Simple single-stage piston (lubricated)

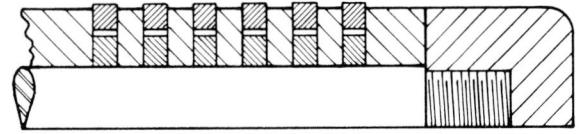

Built-up piston for high-pressure compressor

Figure 16.3 Pistons for air and gas compressors

Pistons 16

GASOLINE ENGINE PISTONS

The key features of a piston for a gasoline engine are shown in Figure 16.4.

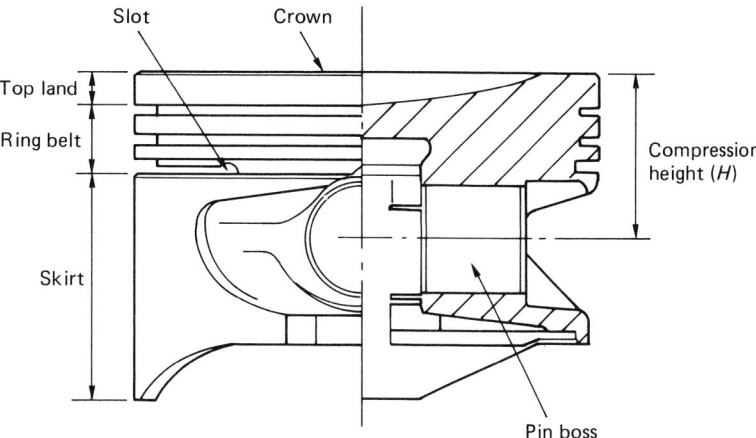

Figure 16.4 Features of a gasoline engine piston

The skirt guides the piston and must be machined slightly out-of-round to compensate for the thermal expansion of the piston and to facilitate hydrodynamic lubrication between the piston and the cylinder bore.

In gasoline pistons in engines of less than 65 bhp/litre slots may be used around part of the circumference of the piston to act as a thermal break between the skirt and the crown, and to introduce more flexibility into the skirt. This allows the running clearance of the skirt to be minimised so as to improve piston stability.

The compression height of the piston (H) is governed by the top land height, the ring belt height and the piston pin diameter. The compression height of gasoline pistons is kept to a minimum in order to reduce reciprocating mass and give improved powertrain refinement.

Piston weight may be characterised by the apparent density (K), defined as follows:-

$$K = \frac{M}{D^3} \text{ gcm}^{-3}$$

where M = mass in grams of bare piston (no rings or pin)

D = bore diameter in centimetres.

The relationship between apparent density and compression height for lightweight designs of gasoline pistons is shown in Figure 16.5.

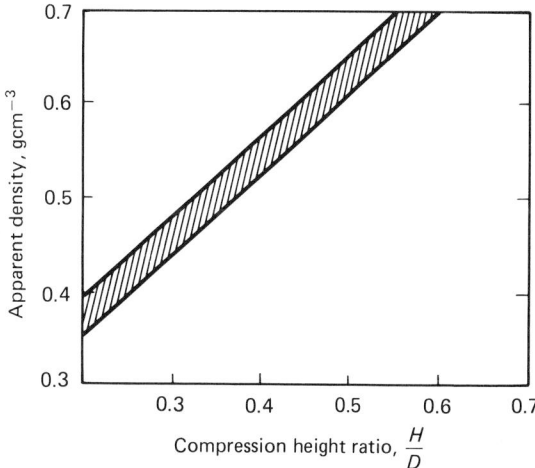

Figure 16.5 The variation of apparent piston density with compression height for a lightweight gasoline piston.

16 Pistons

DIESEL ENGINE PISTONS

The higher loads experienced by diesel pistons means that additional features are commonly necessary. These are shown below:-

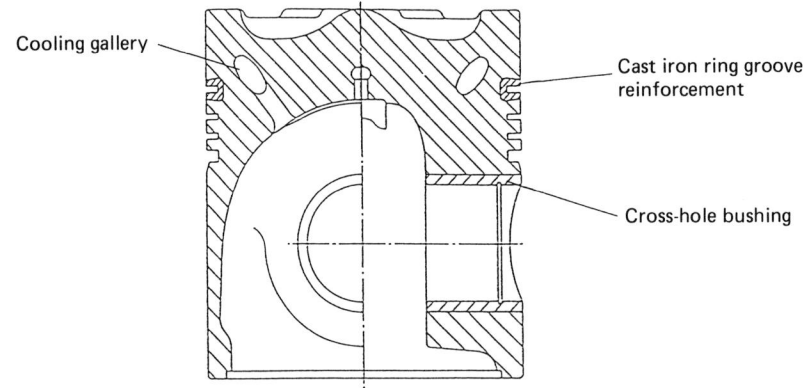

Figure 16.6 A typical highly rated diesel engine piston

The reinforcement of the top ring groove is usually by means of an austenitic cast iron insert, integrally joined to the parent material during the piston casting process.

At engine ratings above 3.5 MW/m^2 of piston area internal oil cooling galleries are often adopted. The gallery is formed by a soluble salt core which is removed after the casting has been produced.

Two-piece pistons are usually specified for cylinder sizes in excess of 300 mm, particularly for engines burning heavy fuel. They are also required at engine ratings above 5.0 MW/m^2 of piston area. The most prevalent combinations are a steel crown with a forged aluminium body and a steel crown with a nodular cast iron body.

Typical bolted crown designs are shown in Figure 16.7.

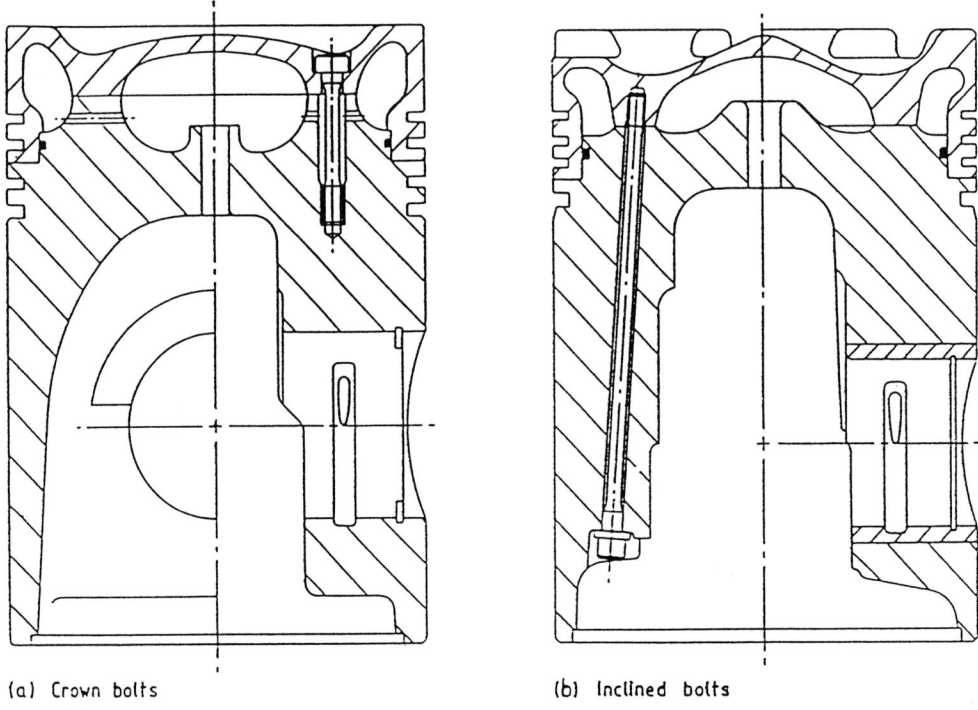

Figure 16.7 Piston designs with bolted crowns

Pistons 16

I.C. ENGINE PISTON DESIGN

Piston materials

The % composition of typical eutectic piston alloys is shown below:-

	Gasoline pistons %	Diesel pistons %
Silicon	10.0–13.0	11.0–12.5
Copper	0.7–1.5	0.7–1.5
Magnesium	0.8–1.5	0.7–1.3
Iron	1.0 max	0.5 max
Manganese	0.5 max	0.25 max
Zinc	0.5 max	0.1 max
Aluminium	remainder	remainder

Ring arrangements

Typical ring arrangements for the four main types of IC engines are shown in Figure 16.8.

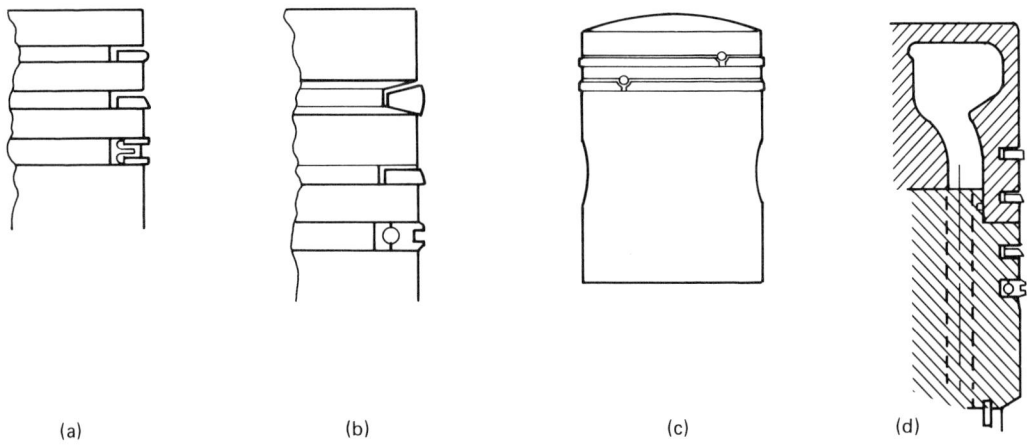

(a) Typical four-stroke petrol engine ring layout:
(b) Typical four-stroke diesel engine ring layout:
(c) Two-stroke petrol engine piston showing ring layout:
(d) Four-stroke medium speed diesel piston showing ring layout

Figure 16.8 Typical ring arrangements for the four main types of IC engines

16 Pistons

Piston dimensions

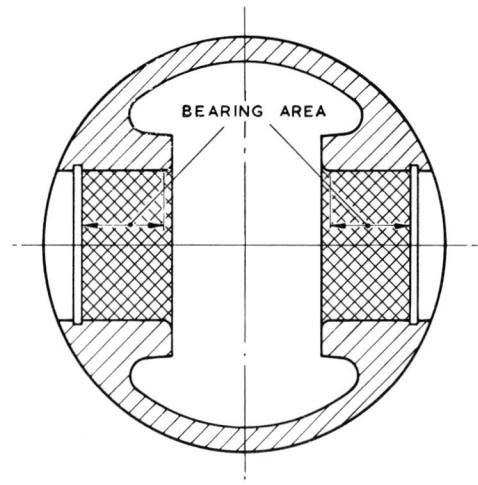

Gudgeon pin bearing area

Gudgeon pin diameter to be decided so that piston boss bearing area will support total pressure loading on piston, with bearing pressure not exceeding 69 MN/m^2 (10 000 lbf/in^2) in aluminium. Gudgeon pin thickness is determined from Figure 16.10 and fatigue strength from Figure 16.11 to ensure that suitable steel is used (DIN 17210 [1984] - 17 Cr 3 or 16 Mn Cr 5 where higher fatigue strength required).

Minimum ring side clearance

	DIESEL	PETROL
Top Ring	0.064 mm	0.04 mm
2nd Ring	0.050 mm	0.03 mm
Oil Ring	0.038 mm	0.038 mm

Minimum intermediate land thickness: calculated from ring groove depth and operating pressure – see Figure 16.9

NB: For turbo-charged designs increase calculated width by 29%

Minimum piston/bore clearances: internal combustion engines per each mm of cylinder bore diameter.

Piston dimensions

	Solid skirt mm	Top land mm
Lo-Ex silicon aluminium alloy (LM13 type)	0.00025	0.006

Surface finishes on wearing surfaces	μm Ra
Ring groove side faces	0.6 max.
Skirt bearing surface (aluminium-diamond turn)	3.2 – 4.8
Gudgeon pin o.d. 80 mm	0.1 max.
Gudgeon pin o.d. 80 – 140 mm	0.2 max.
Gudgeon pin bore in piston	0.3 max.

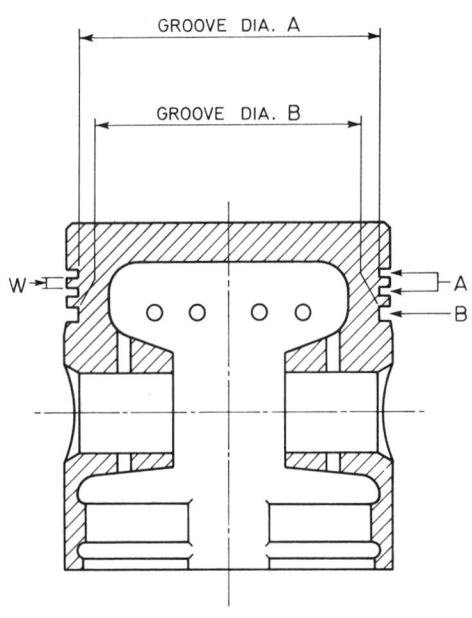

r = MAXIMUM GROOVE ROOT FILLET RADIUS
D = NOMINAL CYLINDER BORE DIA.
t = MAXIMUM RING RADIAL DEPTH

DIA. A = $D - (2t + 0.006\,D + 0.2 + 2r)$ mm
DIA. B = $D - 2(t + r)$ mm

Pistons 16

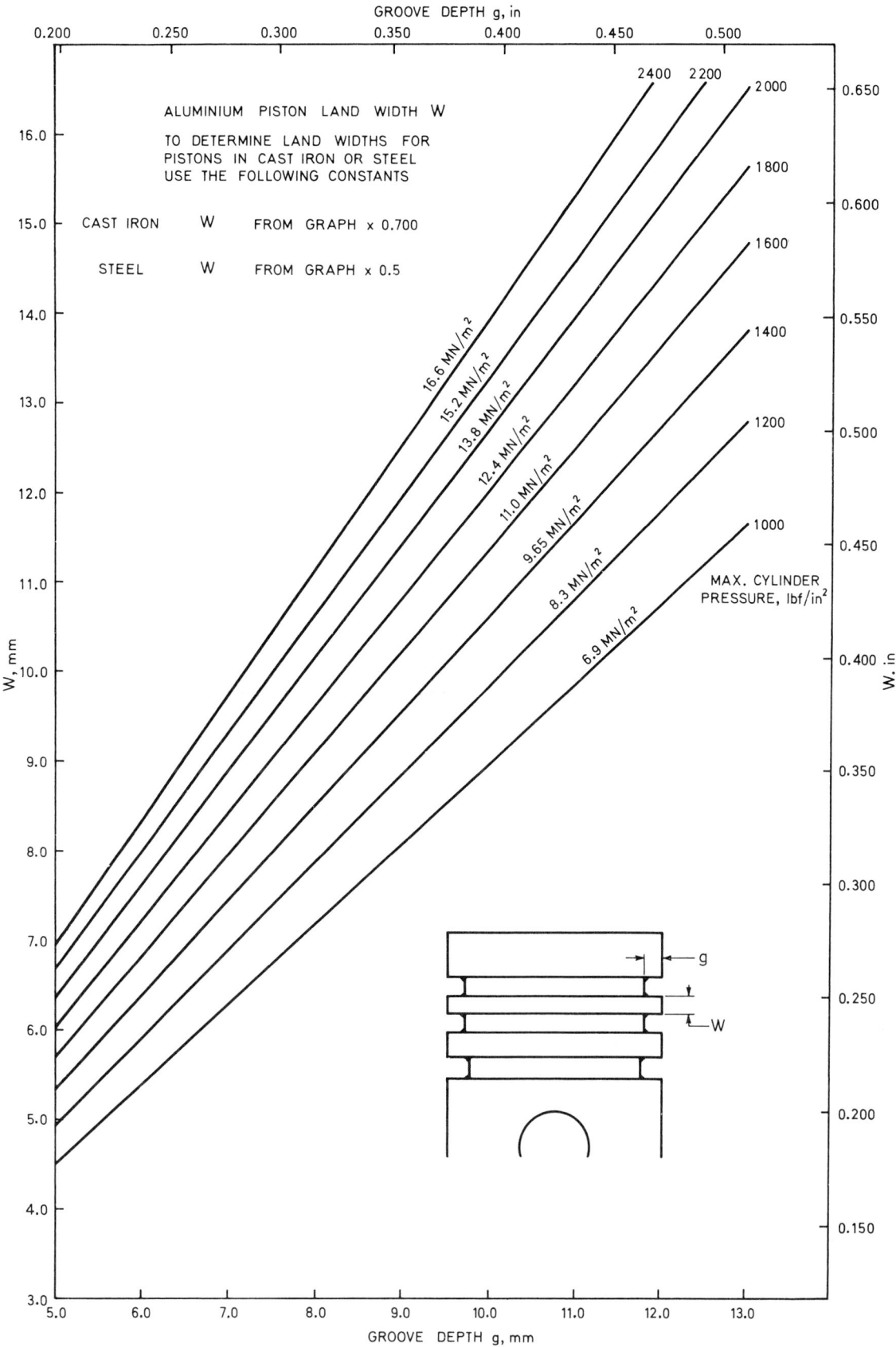

Figure 16.9 The land width required for various groove depths and maximum cylinder pressures

16 Pistons

Gudgeon pin dimensions

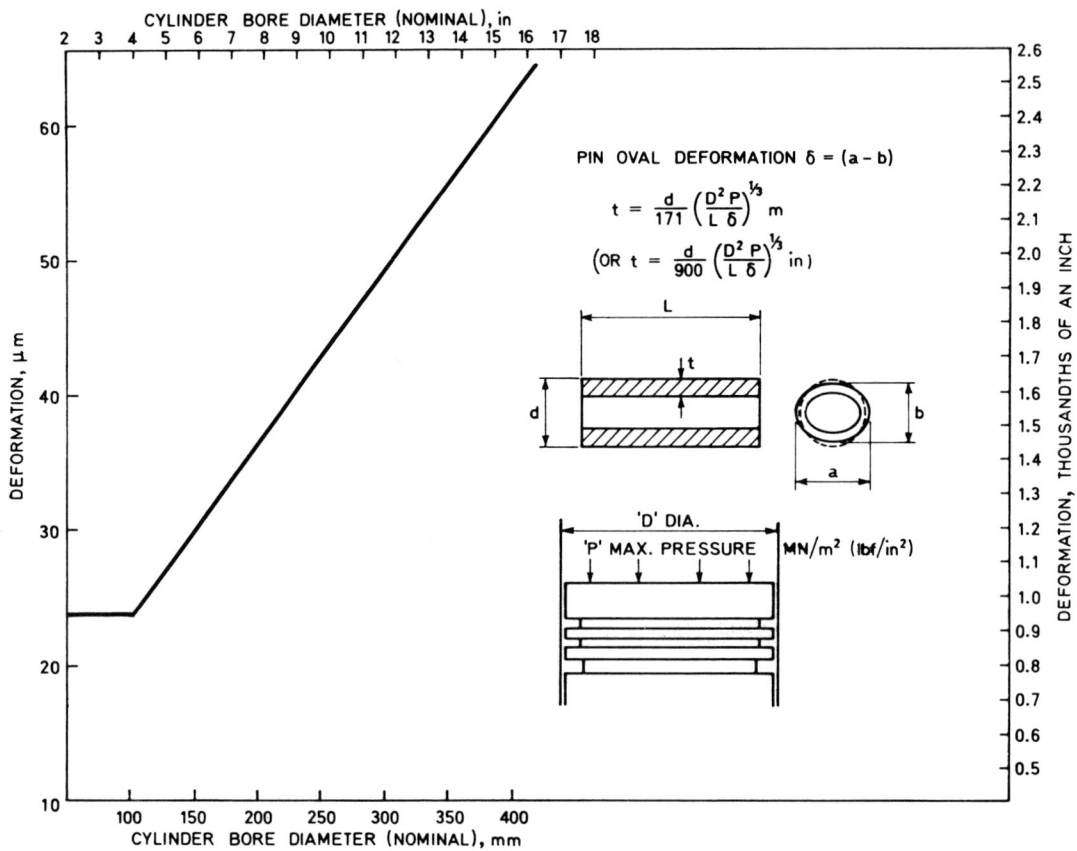

Figure 16.10 Gudgeon pin allowable oval deformation

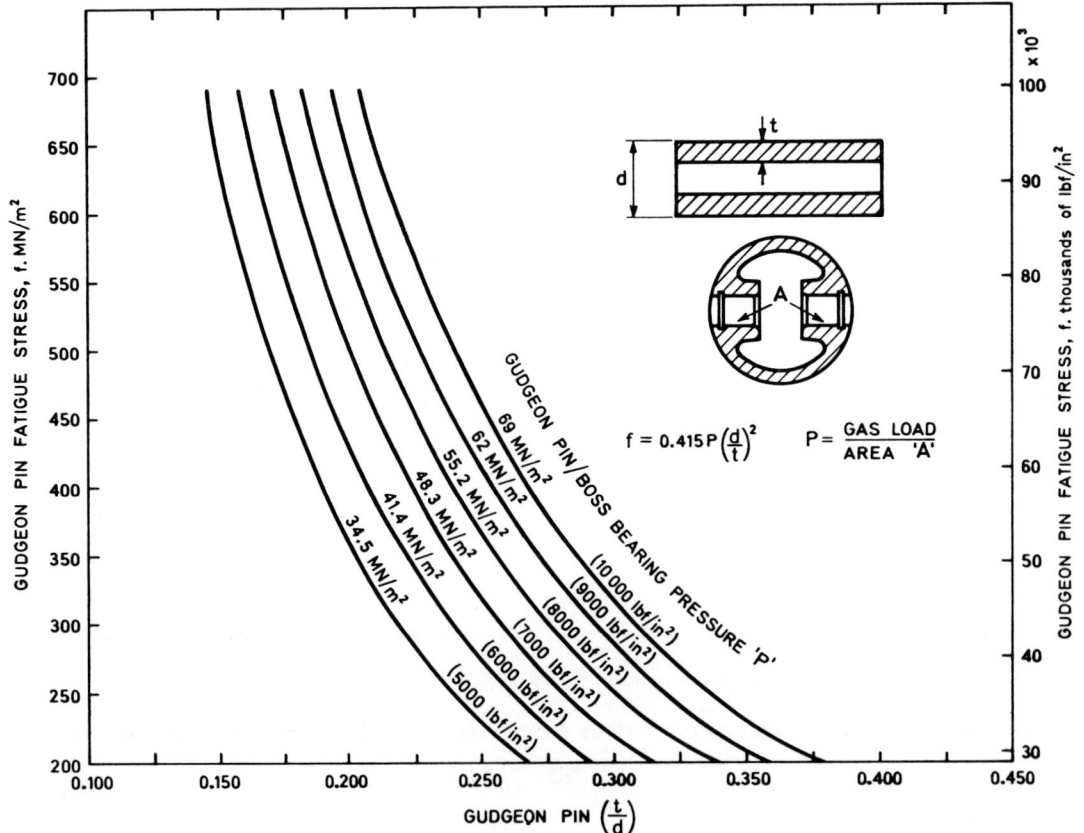

Figure 16.11 Fatigue stress in gudgeon pins for various pin and piston geometries

METALLIC LUBRICATED RINGS

Selection of type and materials

Table 17.1 Types of piston rings and guidance on selection*

Group	Ring type		Description
COMPRESSION RINGS	STRAIGHT-FACED RECTANGULAR	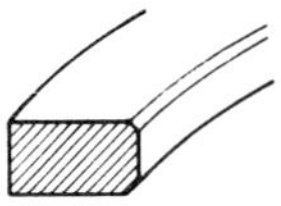 1	The most simple ring can be chromium-plated on peripheral face to give longer life
COMPRESSION RINGS	BARREL-FACED CHROMIUM-PLATED	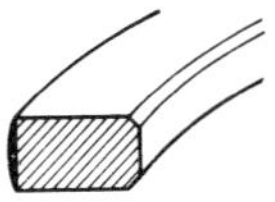 2	Widely used as top compression ring in diesel and petrol engines. Gives quick bed-in, good scuff resistance and long life. Has neutral oil control characteristic
COMPRESSION RINGS	RECTANGULAR INLAID	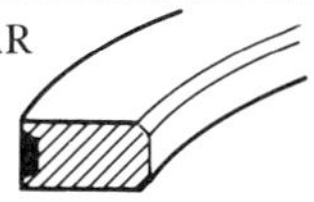 3a	Various low-wear-rate scuff-resistant materials, such as electroplated chrome, sprayed chrome, molybdenum, etc. are set into ring periphery. Outer lands give edge protection to deposited material
COMPRESSION RINGS	SEMI-INLAID	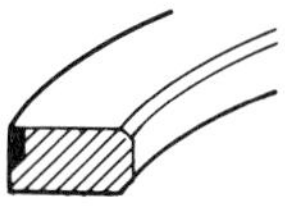 3b	
COMPRESSION RINGS	KEYSTONE	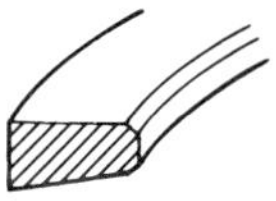 4a	A common top ring in diesel engines prevents sticking due to carbon formation fitted in groove with similar taper
COMPRESSION RINGS	HALF KEYSTONE	 4b	
COMPRESSION RINGS	TAPER FACED	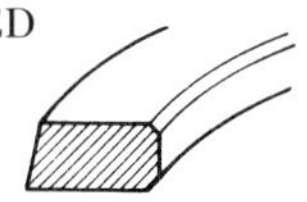 5	Normally between $\frac{1}{2}^\circ$ to $1\frac{1}{2}^\circ$, gives quick bed-in and combines gas sealing and oil control features. A witness land at periphery is often added
COMPRESSION RINGS	DYKES PRESSURE BACKED	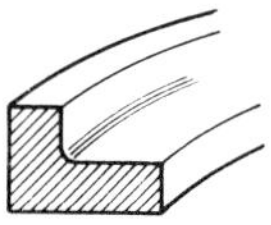 6	Mainly used in high-speed racing applications to prevent blow-by due to 'flutter' under high inertia loading. Fitted in a groove of similar shape
COMPRESSION RINGS	INTERNALLY STEPPED POSITIVE TWIST TYPE	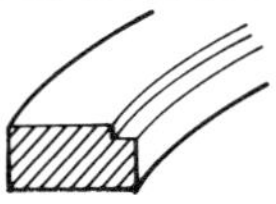 7a	Step or bevel relief on inner edge causes ring to dish when fitted, giving bottom edge contact and good oil control
COMPRESSION RINGS	NEGATIVE TWIST TYPE	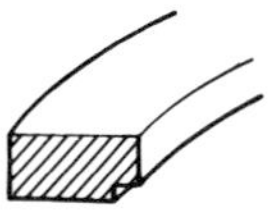 7b	
COMPRESSION RINGS	GROOVED INLAID	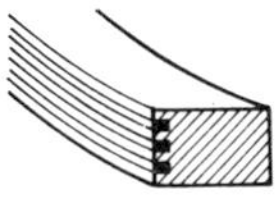 8	Materials such as chrome, bronze and ferrox are inlaid in multi-groove configurations, providing good scuff resistance
DUAL PURPOSE	EXTERNALLY-STEPPED COMPRESSION AND SCRAPER	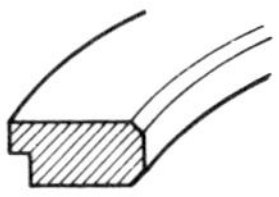 9	Combines gas sealing and oil control functions, the step giving the ring a torsional twist when fitted
DUAL PURPOSE	NAPIER	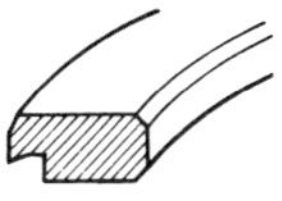 10	Variation of externally stepped ring; hooked relief gives sharp scraping edge with good oil control
OIL CONTROL RINGS	COIL SPRING LOADED SLOTTED OIL CONTROL	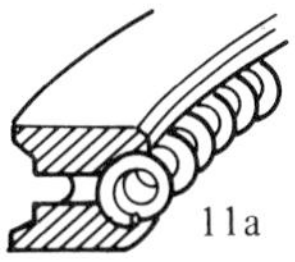 11a	Very popular ring particularly in high-speed diesels. Main wall pressure is derived from butting helical coil expander. Chromium-plated for wall pressures above 700 kN/m^2
OIL CONTROL RINGS	BEVELLED EDGE TYPE	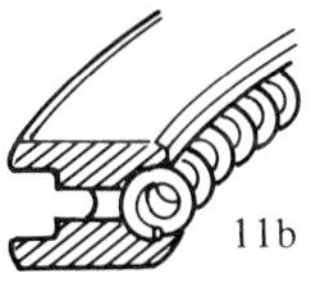 11b	
OIL CONTROL RINGS	SLOTTED OIL CONTROL	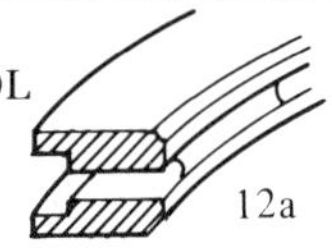 12a	A common form of bulk oil scraper with two scraping lands separated by drainage slots or holes
OIL CONTROL RINGS	BEVELLED EDGE TYPE	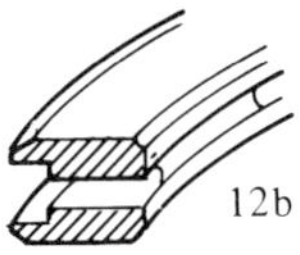 12b	
OIL CONTROL RINGS	STEEL RAIL MULTIPIECE (*a*)	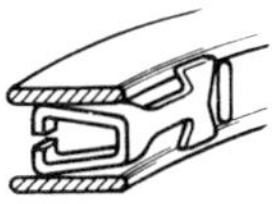	(*a*) Combined spacer expander (*b*) Separate spacer expander These rings allow very high scraping pressures to be applied with good conformability, and have chromium-plated rails
OIL CONTROL RINGS	(*b*)	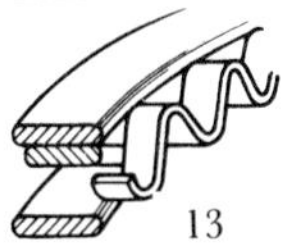 13	

* Basic ring types only shown above. Several features often combined in one ring design, e.g. 4, 7 and 2, giving internally stepped, barrel-faced, chromium-plated, keystone ring.

17 Piston rings

Table 17.2 Properties of typical ring materials

	Modulus of elasticity E_n GN/m^2	Tensile strength MN/m^2	Hardness BHN	Fatigue rating	Wear rating	Scuff compatibility rating
Grey irons	83–124	230–310	210/310	Fair/good	Good	Very good/excellent. Good on chrome
Carbidic malleable irons	140–160	400–580	250/320	Good/very good	Excellent	Good. Very good on chrome
Malleable/nodular irons	155–165	540–820	200/440	Excellent	Poor. Usually chromium-plated	Poor. Usually chromium-plated
Sintered irons	120	250–390	130/150	Good	Good	Very good

Table 17.3 Ring coatings

Coating	Ring wear	Scuff/compatibility	Bore wear	Comment
Chromium—electroplated	Excellent	Very good	Very good	Most widely used coating
Chrome sprayed	Very good	Very good	Fair	Wide variation in bore wear performance
Molybdenum sprayed	Fair	Excellent	Fair	Suffers from temperature/time break-up, especially above 250°C
Tungsten-carbide sprayed	Excellent	Good	Good	—
Iron oxide (ferrox)	Fair	Very good	Good	—
Phosphates (Parko-lubrising)	Fair	Running-in scuff very good	—	Mainly used for running-in
Copper plating	Poor	Running-in scuff excellent	Very good	Mainly used for running-in. Can also be applied to chromium-plate

Ring design and performance

Ring design uses well established elastic bending theory and is based on a careful compromise between opening stress when fitting the ring on to the piston, and working stress when the ring is in the cylinder.

The uniform elastic wall pressure P is given by:

$$P = \frac{E_n L}{7.07 D (D/t - 1)^3}$$

where E_n = nominal modulus of elasticity
L = free gap
D = external ring diameter
t = radial thickness.

Depending on application, the ring free shape is sometimes modified to give high or low pressure at the horns. This is termed positive or negative ovality and is measured by means of a flexible band placed round the ring and closed to the bore diameter (Figure 17.1). Typically, negative ovality is used in two-stroke applications where ports have to be crossed and highly rated top ring applications where compensation is required to offset thermal distortion effects. Positive ovality limits the onset of flutter in high-speed applications.

To obtain a uniform pressure distribution, the ring free shape is given by:

Radial ordinate

$$R_c = R + U + \delta u$$

where

$$U = \frac{FR^4}{E_n I} (1 - \cos \alpha + \tfrac{1}{2} \alpha \sin \alpha)$$

$$\delta u = \frac{R}{2} \left(\frac{FR^3}{E_n I} \right)^2 (\alpha - \tfrac{1}{2} \alpha \cos \alpha - \tfrac{1}{2} \sin \alpha)(3 \sin \alpha + \alpha \cos \alpha)$$

F = mean wall pressure × ring axial width
R = radius at neutral axis, when in the cylinder
I = moment of inertia

(see Figure 17.2)

Piston rings 17

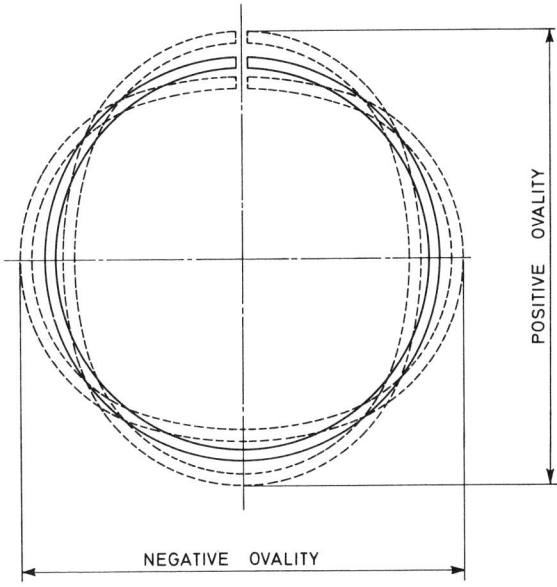

Figure 17.1 Definition of positive and negative ovality

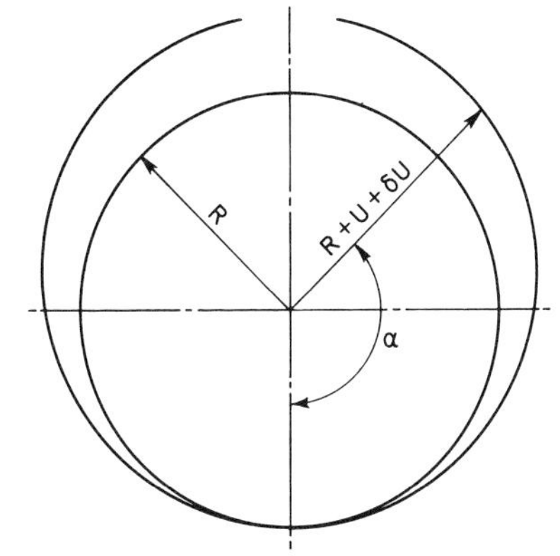

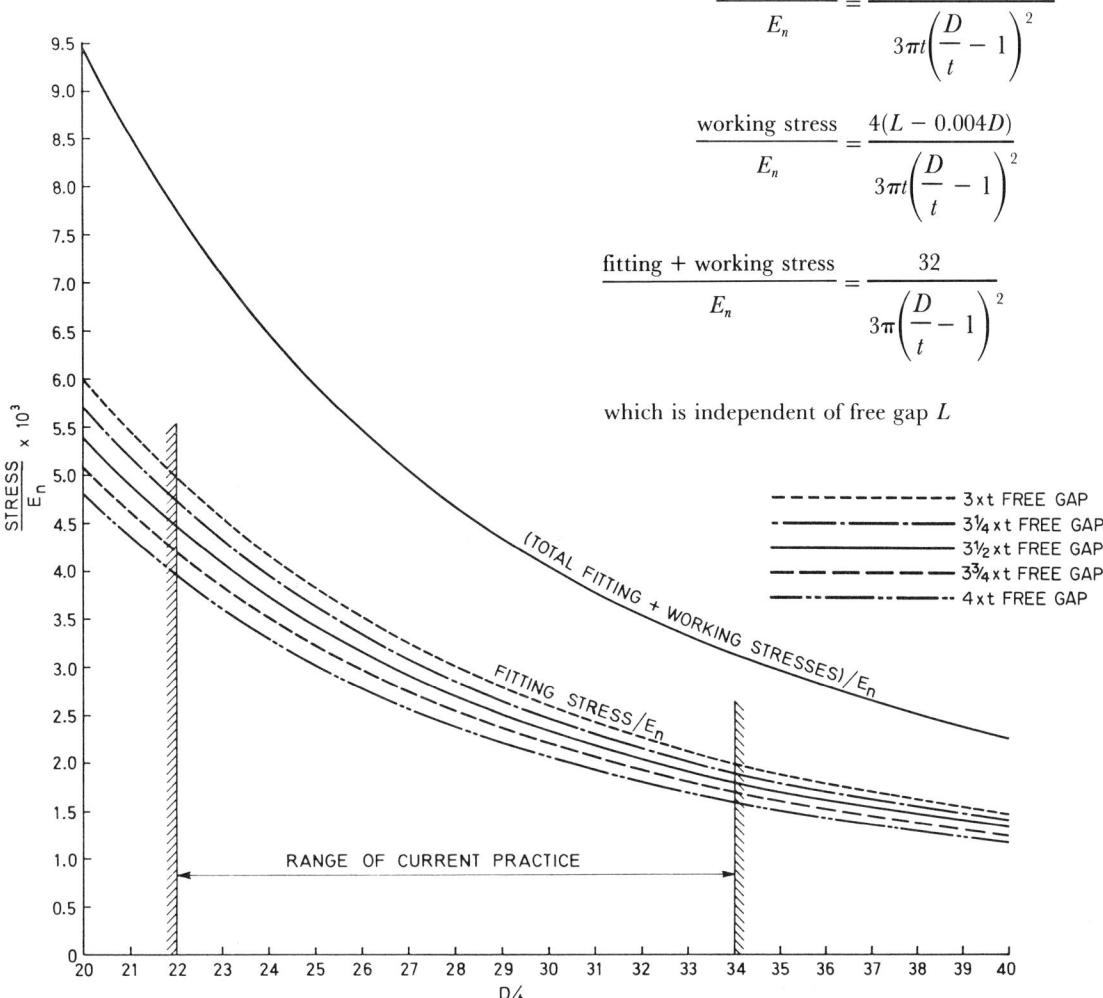

Figure 17.2 Relation between free and closed ring shape for constant pressure distribution

$$\frac{\text{fitting stress}}{E_n} = \frac{4(8t - L + 0.004D)}{3\pi t \left(\dfrac{D}{t} - 1\right)^2}$$

$$\frac{\text{working stress}}{E_n} = \frac{4(L - 0.004D)}{3\pi t \left(\dfrac{D}{t} - 1\right)^2}$$

$$\frac{\text{fitting + working stress}}{E_n} = \frac{32}{3\pi \left(\dfrac{D}{t} - 1\right)^2}$$

which is independent of free gap L

Figure 17.3 The effect of the ring radial thickness on the fitting and working stresses

17 Piston rings

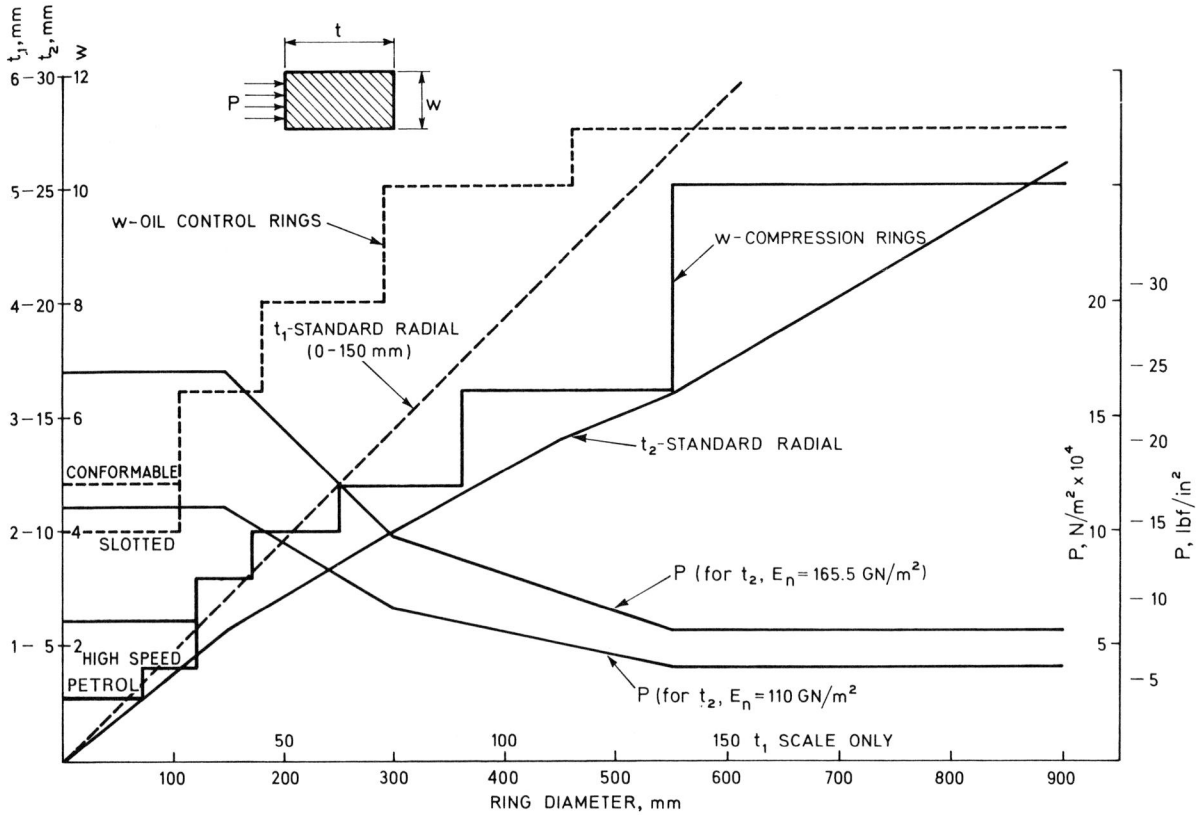

Figure 17.4 Typical ring dimensions

Table 17.4 Minimum side clearance

Application	Bore size (mm)	Minimum groove side clearance (mm) (on dimension W – Figure 17.4)	
		Top compression ring	2nd and 3rd compression rings Oil control rings
Petrol	50–100	0.035	0.035
	100–150	0.06	0.06
Diesel	75–175	0.06	0.04
	175–250	0.08	0.06
	250–400	0.10	0.08
	400–600	0.15	0.13
	Over 600	0.15	0.13
Compressors	50–150	0.025	0.025
	150–300	0.04	0.04
	300–450	0.05	0.05

Piston rings 17

Ring pack arrangements

Narrower rings are commonly used in gasoline engines. A typical gasoline ring pack is shown in Figure 17.5.

The top ring may be manufactured from steel, and may be surface hardened by a nitriding process, or may be chrome plated or molybdenum-inlaid.

In gasoline engines of less than 2.0 litres, the oil-control ring may be of a multi-piece design.

Typical steel specifications are as follows:-

Nitriding steel – DIN X65 CrMo14

Hardened tempered steel – SAE 9254

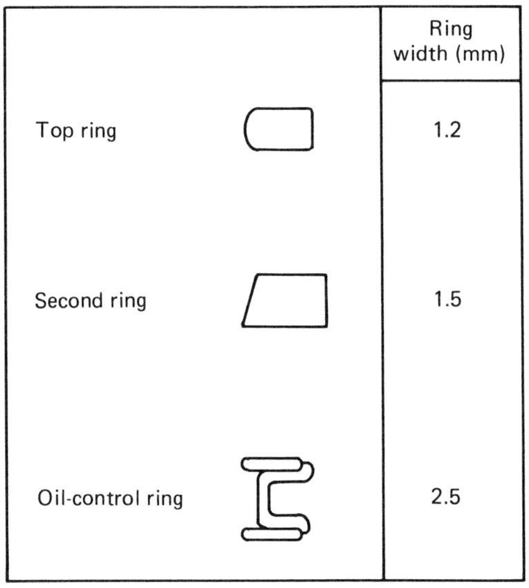

		Ring width (mm)
Top ring		1.2
Second ring		1.5
Oil-control ring		2.5

Figure 17.5 A typical gasoline engine ring pack

In both high and medium speed four-stroke diesel engines, three rings are now common, and typical ring packs are as shown.

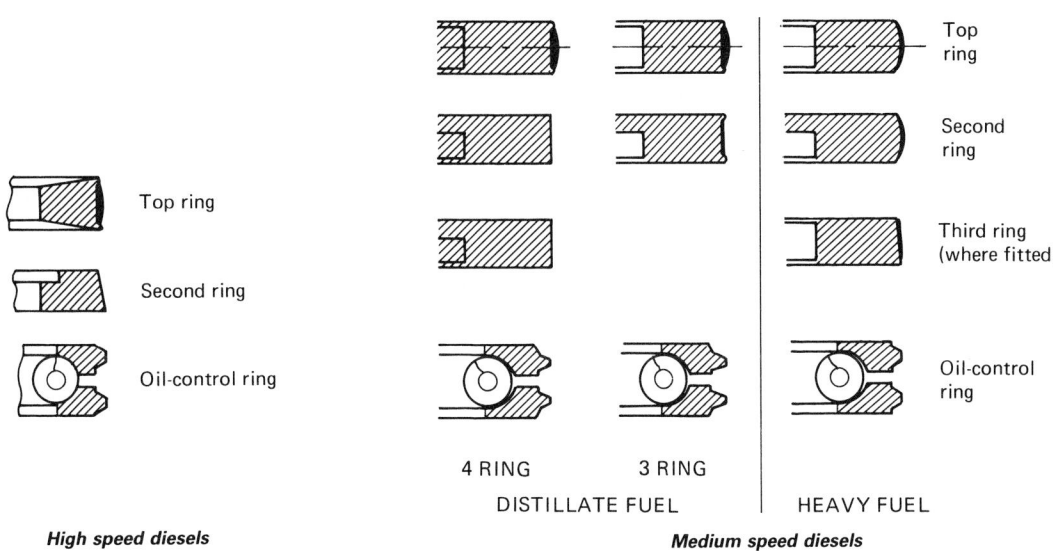

Figure 17.6 Typical ring pack arrangements on high speed and medium speed diesel engines

17 Piston rings

NON METALLIC PISTON RINGS

Metallic piston rings require lubrication for satisfactory operation. There are, however, many applications where lubricants would be considered a contaminant or even a fire hazard, e.g. in food-processing equipment. For these applications, piston rings can be made from self-lubricating materials. These materials can also be used in lubricated applications where there is a risk of lubricant breakdown.

Ring materials

Table 17.5 Typical properties of ring materials

Material	Tensile strength MN/m^2	Specific gravity	Typical coefficients of expansion $\times 10^{-6}$/°C
Carbon-filled PTFE	10	2.05	55
Glass-filled PTFE	17	2.26	80
Graphite/MoS$_2$ filled PTFE	20	2.20	115
Bronze-filled PTFE	13	3.90	118
Resin-bonded PTFE	29	1.75	30
Resin-bonded fabric	110	1.36	22.5/87.5*
Carbon	43	1.8	43
Resin-bonded carbon	20	1.9	20

* Material is anisotropic, thus the lower expansion is parallel to, and the alternative figure is normal to, the plane of pressing.

Table 17.6 Suggested operating conditions for various materials

Material		Terminal pressure bars	Maximum speed m/s	Maximum temp. °C	Average coefficient of friction (dry)	Minimum humidity p.p.m.	Minimal lubrication
Matrix	Filler						
PTFE	Carbon	200	6.0	250	0.1/0.15	3	Very good
	Glass	200	6.0	200	0.1/0.15	3	Very good
	Graphite/MoS$_2$	200	6.0	200	0.12/0.18	3	Very good
	Bronze	100	4.0	200	0.15/0.2	40	Good
Resin-bonded PTFE		200	4.0	200	0.15	10	Very good
Resin-bonded fabric		100	3.0	150	0.15/0.2	40	Poor
Carbon		60	3.8	350	0.2/0.25	40	Poor
Resin-bonded carbon		100	4.5	180	0.2	40	Poor

Ring design

Table 17.7 Preferred number of rings

Differential pressure bars	0–9	10–14	15–24	25–29	30–49	50–99	100–200
Minimum number of rings	2	3	4	5	6	7	8

Piston rings 17

Table 17.8 Suggested sizes of rings

Units	Cylinder diameter (b)	Piston ring		Groove side clearance (C)
		Axial width	Radial thickness	
Millimetres	25–75	3–5	3–6	C = 0.025 × axial width
	76–150	5–10	5–10	
	151–230	6–12	6–12	(N.B.—Groove tolerance H7)
	230–400	10–19	10–19	
	400–800	12–25	12–25	
Inches	1–3	0.125–0.187	0.125–0.218	
	3–6	0.187–0.375	0.187–0.437	
	6–9	0.250–0.500	0.250–0.500	
	9–16	0.375–0.750	0.375–0.750	
	16–30	0.500–1.000	0.500–1.000	

Table 17.9 Types of joints

Butt	Scarf	Lap or step
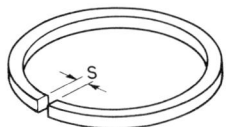	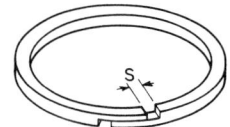	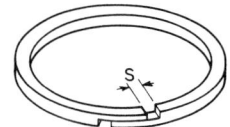
Suitable for all pressures	Suitable for all pressures	Not recommended where pressure differential exceeds 10 atmospheres
Circumferential clearance (S)	$S = \pi \times D \times \alpha_p \times T$ where D = cylinder diameter, α_p = Coeff. of expansion of piston ring material, T = Operating temperature	

Cylinder materials and finishes

Table 17.10 Typical cylinder materials

Material	Type	Remarks
Cast iron	ISO R185 220 grade	Suitable only for continuous operation
Ni-Resist	ISO 2892 AUS101. ASTM A436/1	Preferred to cast iron—less danger of corrosion
Stainless steel	ISO 683/1 316S16. AISI 316	Used for machines where long shutdowns occur
Meehanite cast iron	CR or CRS	Similar to Ni-Resist

A suitable surface finish for these cylinder liners is 0.4 to 0.6 μm R_a or 2.4 to 3.6 μm R_{max}.

18 Cylinders and liners

MATERIALS AND DESIGN

Table 18.1 Choice of materials for cylinders and cylinder liners

Application	Type of construction	Comment	Material — Block	Material — Liner	Comment	Surface finish and treatment
I.C. engines	Monobloc	Most petrol engines. Some oil engines ('siamesed' cylinders used for cheaper and low specific power engines or where space is at a premium)	Grey C.I. (low phosphorus)	—	Simplest and cheapest method of building mass-production engines	Most applications use an untreated cross-hatched honed finish—See Note 3
			Aluminium alloy (high silicon) Aluminium with nickel plate containing silicon carbide particles	—	Gives maximum reduction in weight but poses special problems in material compatibility with mating component, i.e. piston and rings	For greater scuff resistance of C.I. a phosphate treatment can be used. For greater wear resistance bores may be hardened on surface, through- or zone-hardened, or hard chromium-plated on cast iron or steel liner. Surface porosity (by reverse plating) is necessary with chromium plating to give scuff resistance. Porous coatings aid oil retention reducing scuffing and wear. Silicon carbide impregnation can be used to combat bore polishing.
	Dry liner	Oil engines and petrol engines	Grey C.I. (low phosphorus)	Grey C.I. (low-to-medium phosphorus)	Liner normally pressed-in but may be slip fit. Much improved wear. Can pose cooling problems. Used for engine reconditioning in monobloc system	
			Aluminium alloy	Grey C.I. (low-to-medium phosphorus)	Liner normally cast-in or pressed-in	Aluminium alloys (high silicon) require special surface finish to allow free silicon to stand out from the matrix. Nickel plate with silicon carbide particles is the most common solution for aluminium bores. Some cheaper aluminium alloys may be used for 'throw-away' engines. Piston skirts may be electroplated with iron or chromium.
	Wet liner	High-performance petrol and most oil engines	Grey C.I.	Grey C.I. (low-to-medium phosphorus) Grey C.I. with silicon carbide impregnation Austenitic cast iron	Wet liner requirement for long life, good cooling and ease of maintenance	
			Aluminium alloy	Grey C.I. austenitic C.I. Aluminium alloy (high silicon) Aluminium with nickel plate containing silicon carbide particles	High-performance petrol engines to reduce weight	Costs rise significantly from the basic monobloc cast-iron cylinder block. Care must be taken to ensure the minimum specification consistent with technical requirements
Compressors	Monobloc	Small size and low pressure (up to 3–4 in dia. and 100 p.s.i.)	Grey C.I. (low phosphorus)	—	As in i.c. engines	As in i.c. engines except where plastics piston rings are used in which case honed 'mirror-finish' is desirable
	Wet liner	Heavy duty long life high reliability units of all sizes and operating pressures	As in i.c. engines		As in i.c. engines	
Hydraulic actuators and fluid piston pumps	To suit design requirements	—	Grey cast iron Bronze Aluminium alloy Steel		Material depends on environmental requirements of pressure, duty, reliability and fluid in use	Fine turned or honed to mirror finish Hardened steel bores usually ground or lapped

Cylinders and liners 18

Table 18.2 Cylinder/cylinder liner tolerances

	Ovality mm	Concentricity mm
Monoblocs	0.025 FIM max	
Press fit dry type cylinder liners	0.150 FIM max	0.150 FIM max
Slip fit dry type cylinder liners*	0.050 FIM max	0.100 FIM max
Wet type cylinder liners*	0.025 FIM max	0.100 FIM max

* It is also vital that the flange be parallel and square to the major axis of the liner within 0.050 mm.

Table 18.3 Interference fits

Diameter mm	1 Cast iron liners in cast iron blocks 2 Aluminium liners or austenitic iron liners in aluminium blocks mm	Grey cast iron in aluminium blocks mm
Up to 40	0.025 min	0.075 min
over 40 to 50	0.040 min	0.100 min
over 50 to 75	0.050 min	0.115 min
over 75 to 100	0.075 min	0.125 min
over 100 to 155	0.100 min	0.140 min

Notes:

1. Choice of construction and material is dependent on market being catered for: i.e. cost, power output or delivery requirement, life requirement, size and intended application.
2. Choice of material is also dependent on material used for pistons and rings and on any surface coatings given to these. Also, but to a lesser extent, on the surface treatment.
3. Honing specifications generally satisfactory; lies in the range 20 to 40 micro-inches, with a horizontal included angle of cross-hatch of 30/60° and a 60% plateau area. Surfaces must be free from folds, tears, burrs and burnished areas (see illustration). Suitable surface conditions can be most easily accomplished with silicon carbide hones. Diamond hones can be used but are best confined to roughing-cuts. Finishing can then be performed with silicon carbide honing stones or for more critical applications with silicon carbide particles in a soft matrix such as cork. Control of production tools and machines is vital for satisfactory performance in series production.
4. Sealing of wet liners is of great importance—see BS 4518 for Sealing Rings. Proprietary sealant/adhesive materials are available for assisting in sealing and in fixing liners.

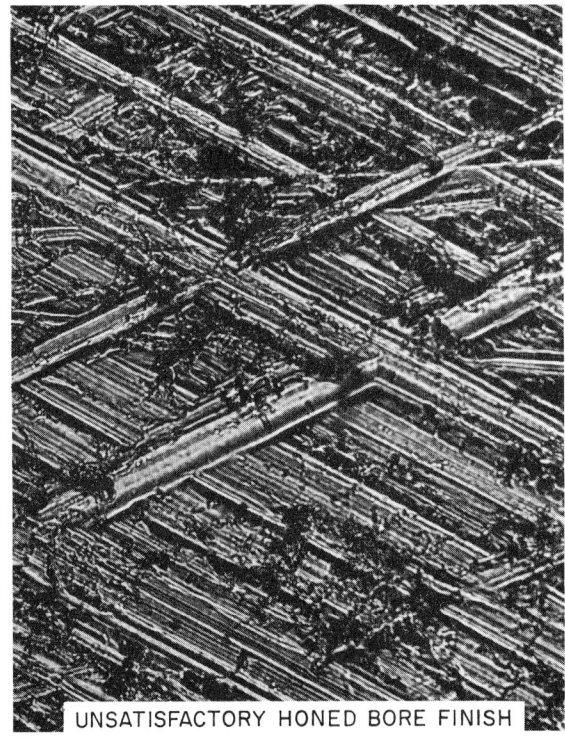

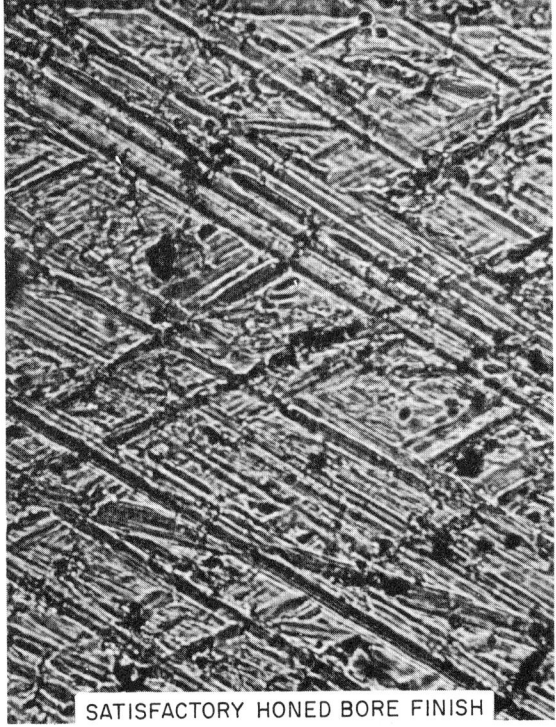

18 Cylinders and liners

Table 18.4 Materials, compositions and properties

Type	Typical composition (%) (Remainder is F_e)								Typical properties		
	C	Si	S	P	Mn	Ni	Cr	Others	Coeff. of thermal expansion $20°$-$200°C$	U.T.S	BHN
Sand-cast blocks and barrels	3.3	2.1	0.1	0.15	0.6	0.3	0.2		$11.0 \times 10^{-6}/°C$.0 ×10	220MN/m^2	200
Sand-cast liners	3.3	1.8	0.1	0.25	0.8	—	0.4		$11.0 \times 10^{-6}/°C$	230MN/m^2	200
Centrifugally-cast grey iron liners	3.4	2.3	0.06	0.5	0.8	—	0.4		$11.0 \times 10^{-6}/°C$	260MN/m^2	250
Centrifugally-cast alloy iron liners	3.3	2.2	0.06	0.2	0.8	—	0.4	Ni and Cu 0.5–1.5 Mo/Va 0.4	$10.5 \times 10^{-6}/°C$	320MN/m^2	280
Austenitic iron liners	2.9	2.0	0.06	0.3	0.8	14.0	2.0	Cu 7.0	$19.3 \times 10^{-6}/°C$	190MN/m^2	180

Table 18.5 Microstructures required

Type	Microstructure
Sand-cast blocks and barrels	Flake graphite, pearlitic matrix, no free carbides, phosphide eutectic network increases with phosphorus content, minimum of free ferrite desirable to minimise possibility of scuffing but less important with increasing phosphide
Sand-cast liners	As for sand cast but with finer graphite tending towards rosette or undercooled. Matrix martensitic/bainitic if linear hardened and tempered
Centrifugally-cast grey iron liners	Compact graphite or quasi-flakes, pearlitic matrix, islands of wear-resistant alloy carbides distributed throughout (approx. 5% by volume) matrix. Phosphide exists as ternary eutectic with carbides. Minimum of free ferrite ideal, but not important in presence of carbides
Centrifugally-cast alloy iron liners	Uniformly distributed fine flake graphite along with some undercooled graphite (ASTM Types A along with some D and E, Sizes 5–8. Fine-grained cored austenite matrix. Complex carbides and ternary phosphide eutectic are present in controlled amounts in a broken, non-continuous network
Austenitic iron liners	

Selection of seals 19

BASIC SEAL TYPES AND THEIR CHARACTERISTICS

Dynamic seal

Sealing takes place between surfaces in sliding contact or narrowly separated.

Static seal

Sealing takes place between surfaces which do not move relative to each other.

Pseudo-static seal

Limited relative motion is possible at the sealing surfaces, or the seal itself allows limited motion; e.g. swivel couplings for pipes, flexible diaphragms.

Exclusion seal

A device to restrict access of dirt, etc., to a system, often used in conjunction with a dynamic seal.

Table 19.1 Characteristics of dynamic seals

	Contact seals		Clearance seals
Sealing interface	Surfaces loaded together: (i) Hydrodynamic operation (normal loads, speeds and viscosities) — ABOUT 1 μm — FLUID FILM (a)	(ii) Boundary lubrication (high loads, low speeds, low viscosities) — MOLECULAR FILM (b)	Predetermined separation — ABOUT 25 μm — PRESET GAP (c)
Leakage	(i) Low to very low or virtually zero	(ii) As (i)	High, except for viscoseal and centrifugal seal at design optimum
Friction	Moderate	High	Low
Life	Moderate to good	Short	Indefinite
Reliability	Moderate to good	Poor	Good

Table 19.2 Types of dynamic and static seals

Dynamic seals				Static seals
Contact seals		Clearance seals		
Rotary	Reciprocatory oscillatory	Rotary	Reciprocatory	
Lip seal (Figure 19.1) Mechanical seal (Figure 19.2) Packed gland (Figure 19.3) 'O' ring* Felt ring	'U' ring, etc. (Figure 19.4) Chevron (Figure 19.5) 'O' ring (Figure 19.6) Lobed 'O' ring (Figure 19.7) Coaxial PTFE seal (Figure 19.8) Packed gland (Figure 19.3) Piston ring Bellows (Figure 19.9) Diaphragm	Labyrinth† (Figure 19.10a) Viscoseal (Figure 19.10b) Fixed bushing (Figure 19.10d) Floating bushing (Figure 19.10e) Centrifugal seal (Figure 19.10c) Polymeric bushing (Figure 19.10f)	Labyrinth† (Figure 19.10a) Fixed bushing (Figure 19.10d) Floating bushing (Figure 19.10e)	Bonded fibre sheet Spiral wound gasket Elastomeric gasket Plastic gasket Sealant, setting Sealant, non-setting 'O' ring Inflatable gasket Pipe coupling Bellows

* Only for very slow speeds.
† Usually for steam or gas.

19 Selection of seals

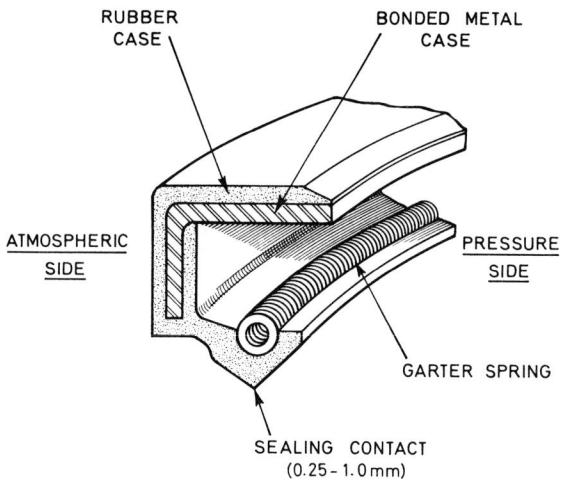

Figure 19.1 Rotary lip seal

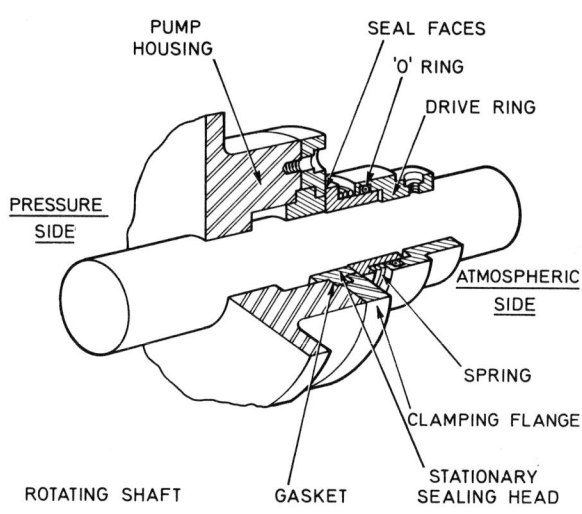

Figure 19.2 Mechanical seal

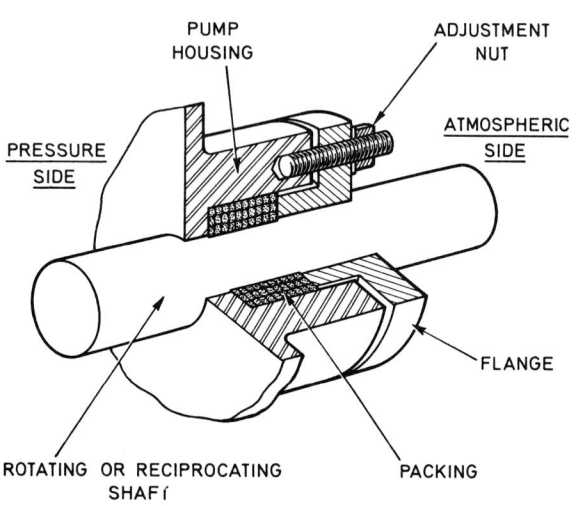

Figure 19.3 Packed gland

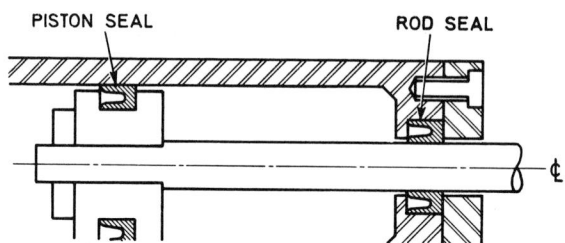

Figure 19.4 Square-backed 'U' seals as piston and rod seals in a hydraulic cylinder

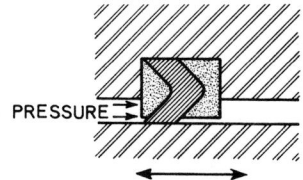

Figure 19.5 Chevron seal with shaped support rings

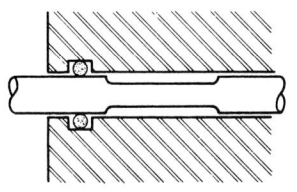

Figure 19.6 'O' ring seal on control valve spool

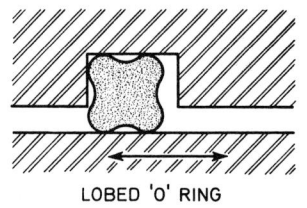

Figure 19.7 Lobed 'O' ring

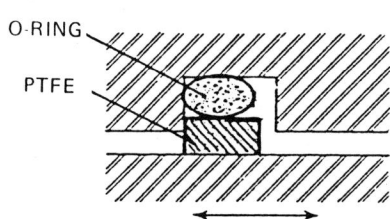

Figure 19.8 Coaxial PTFE seal

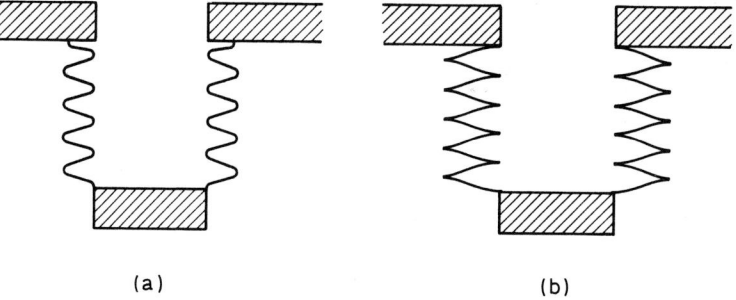

Figure 19.9 Metal bellows: (a) formed; (b) welded

Selection of seals 19

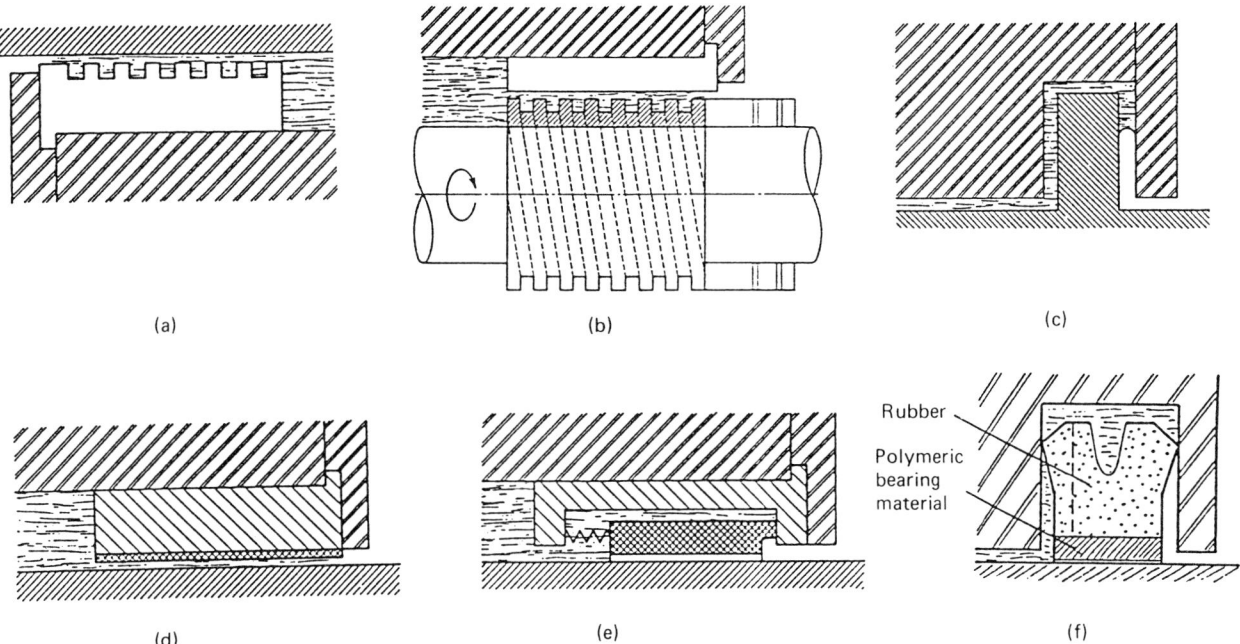

Figure 19.10 Examples of clearance seals: (a) labyrinth; (b) viscoseal; (c) centrifugal seal; (d) fixed bushing; (e) floating bushing; (f) polymeric-bushing seal

Multiple seals

One seal or several in series may be used, depending on the severity of the application. Table 19.3 shows six basic dynamic sealing problems where two fluids have to be separated. Since contact seals rely on the sealed fluids for lubrication of the sliding parts it is essential that the seal(s) chosen should be exposed to a suitable lubricating liquid. Where this is not already so, a second seal enclosing a suitable 'buffer' liquid must be used. Multiple seals are also used where the pressure is so large that it must be broken down in stages to comply with the pressure limits of the individual seals, or where severe limitations on contamination exist. Table 19.3 lists the procedures for dealing with these various situations. Where a buffer fluid is used, care should be taken to ensure proper pressure control, especially when exposed to temperature variation. The pressure drop across successive seals will not be identical unless positive control is provided.

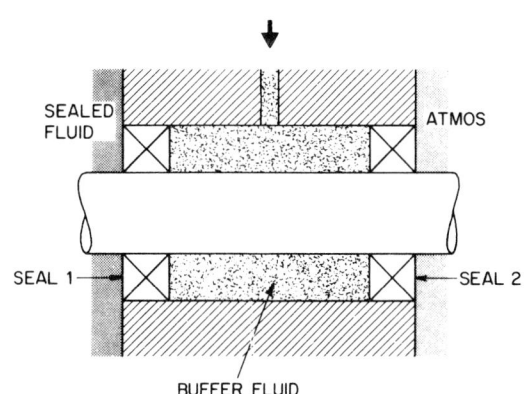

Figure 19.11 Multiple seals, with buffer fluid

Terminology:

'Tandem seals'   multiple seals facing same direction, used to stage the pressure drop of the system. Inter-stage pressures progressively lower than sealed pressure.

'Double seals'   pair of seals facing opposite directions, used to control escape of hazardous or toxic sealed fluid to environment, or to permit liquid lubrication of the inner seal. The buffer pressure is normally higher than the sealed pressure.

19 Selection of seals

SEAL SELECTION

Table 19.3 *The use of dynamic contact seals in the six dynamic sealing situations*

Configuration (see diagram)	Single seal	Multiple seal	

(a)	Satisfactory unless:
	    (i) no contamination permissible
	    (ii) $\|p_1 - p_2\|$ large
	    (iii) liquids both poor lubricants
	    (iv) abrasive present
		Buffer fluid = gas or vacuum:
		 Buffer fluid = liquid 1 or 2:
		 Buffer fluid = good lubricant:
		 Buffer fluid = clean liquid:
			$p_B > p_1, p_2$ or $p_B \ll p_1, p_2$
			 $p_B \simeq (p_1 + p_2)/2$
			 $p_B > p_1, p_2$
			 $p_B > p_1$ or p_2, subject to abrasive location

(b)	Satisfactory unless:
	    (i) no contamination permissible
	    (ii) $\|p_1 - p_2\|$ large
	    (iii) the liquid is a poor lubricant
	    (iv) abrasive present
		Buffer fluid = gas or vacuum:
		
		 Buffer fluid = liquid:
		 Buffer fluid = good lubricant:
		 Buffer fluid = clean liquid:
			$p_B > p_1$ or $p_B > p_2$ or $p_B > p_1, p_2$ or $p_B \ll p_1, p_2$
			 $p_B \simeq (p_1 + p_2)/2$
			 $p_B > p_1, p_2$
			 $p_B > p_1$ or p_2, subject to abrasive location

(c)	Satisfactory unless:
	    (i) no contamination of vacuum permissible
	    (ii) $p_2 \gg p_2$
	    (iii) the liquid is a poor lubricant
	    (iv) abrasive present in liquid
		Buffer fluid = compatible liquid or gas; alternatively use clearance seal(s) and evacuate buffer zone
		
		 Buffer fluid = clean liquid:
			$p_B \gtrsim p_1$
			
			 $p_B > p_1$

(d)	Unsatisfactory	Buffer fluid = compatible liquid lubricant:	$p_B > p_1, p_2$
(e)	Unsatisfactory	Buffer fluid = compatible liquid lubricant:	$p_B > p_1$
(f)	Unsatisfactory	Buffer liquid = compatible liquid lubricant:	$p_B > p_1, p_2$

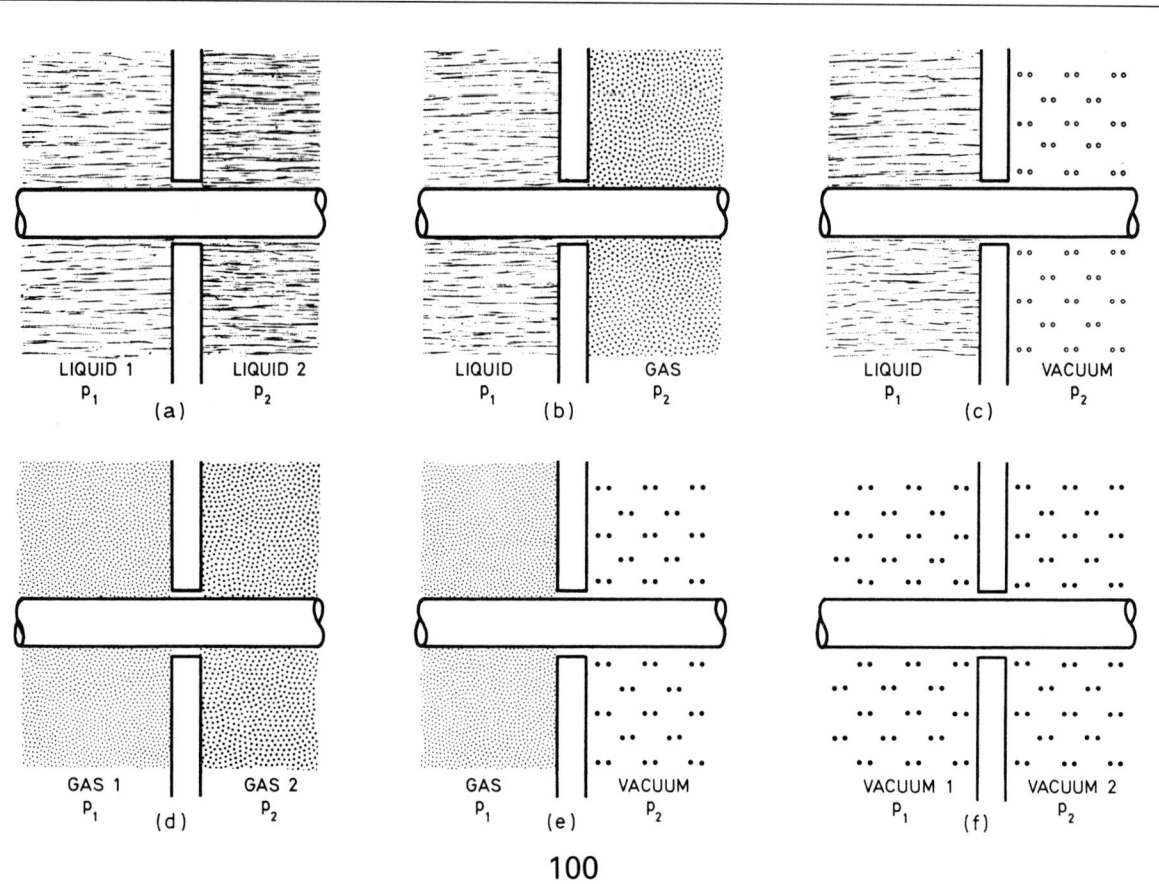

Selection of seals 19

Check-list for seal selection

Temperature (see Figure 19.12): seals containing rubber, natural fibres or plastic (which includes many face seals) may have severe temperature limitations, depending on the material, for example:

Natural rubber	−50 to +80°C
Nitrile rubber	−40 to +130
Fluorocarbon rubber	−40 to +200
Perfluorocarbon rubber	−10 to +300
PTFE, plastic	−100 to +280

At low temperatures, certain of the fluoroelastomers may become less 'rubbery' and may seal less well at high pressure.

Speed (see Figure 19.13)
Pressure (see Figure 19.13)
Size (see Figure 19.14)
Leakage (see Figure 19.15)

After making an initial choice of a suitable type of seal, the section of this handbook which relates to that type of seal should be studied. Discussion with seal manufacturers is also recommended.

Fluid compatibility: check all materials which may be exposed to the fluid, especially rubbers.

Abrasion resistance: harder sliding contact materials are usually better but it is preferable to keep abrasives away from the seal if at all possible, for example by flushing with a clean fluid.

Polyurethane and natural rubber are particularly abrasion resistant polymers. Where low friction is also necessary filled PTFE may be considered.

Vibration: should be minimised, but rubber seals are likely to function better than hard seals.

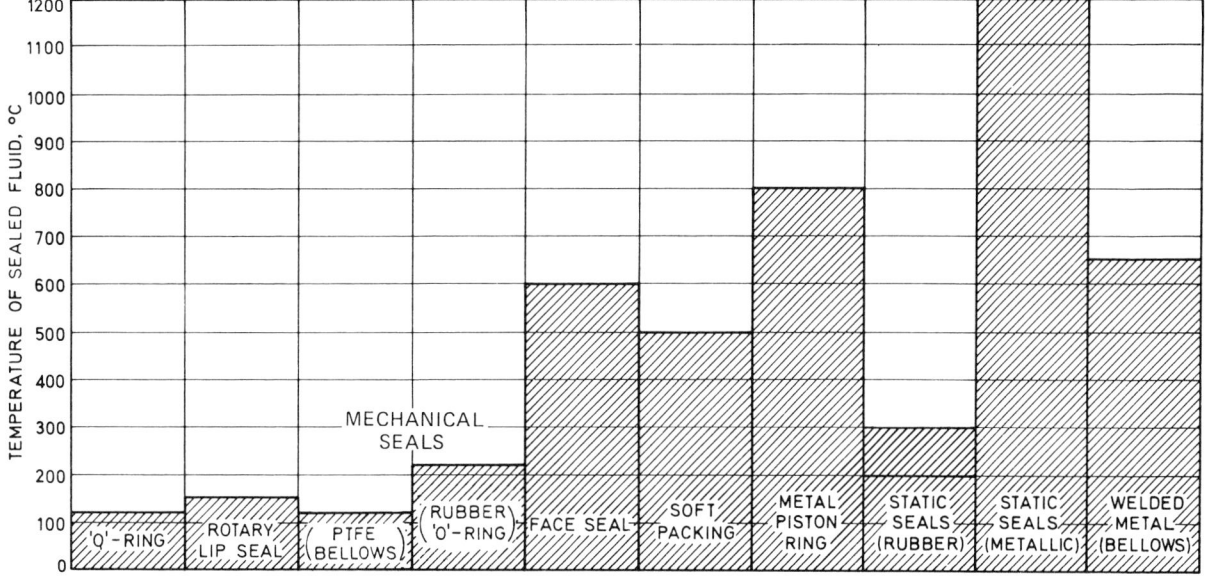

Figure 19.12 Approximate upper temperature limits for seals

19 Selection of seals

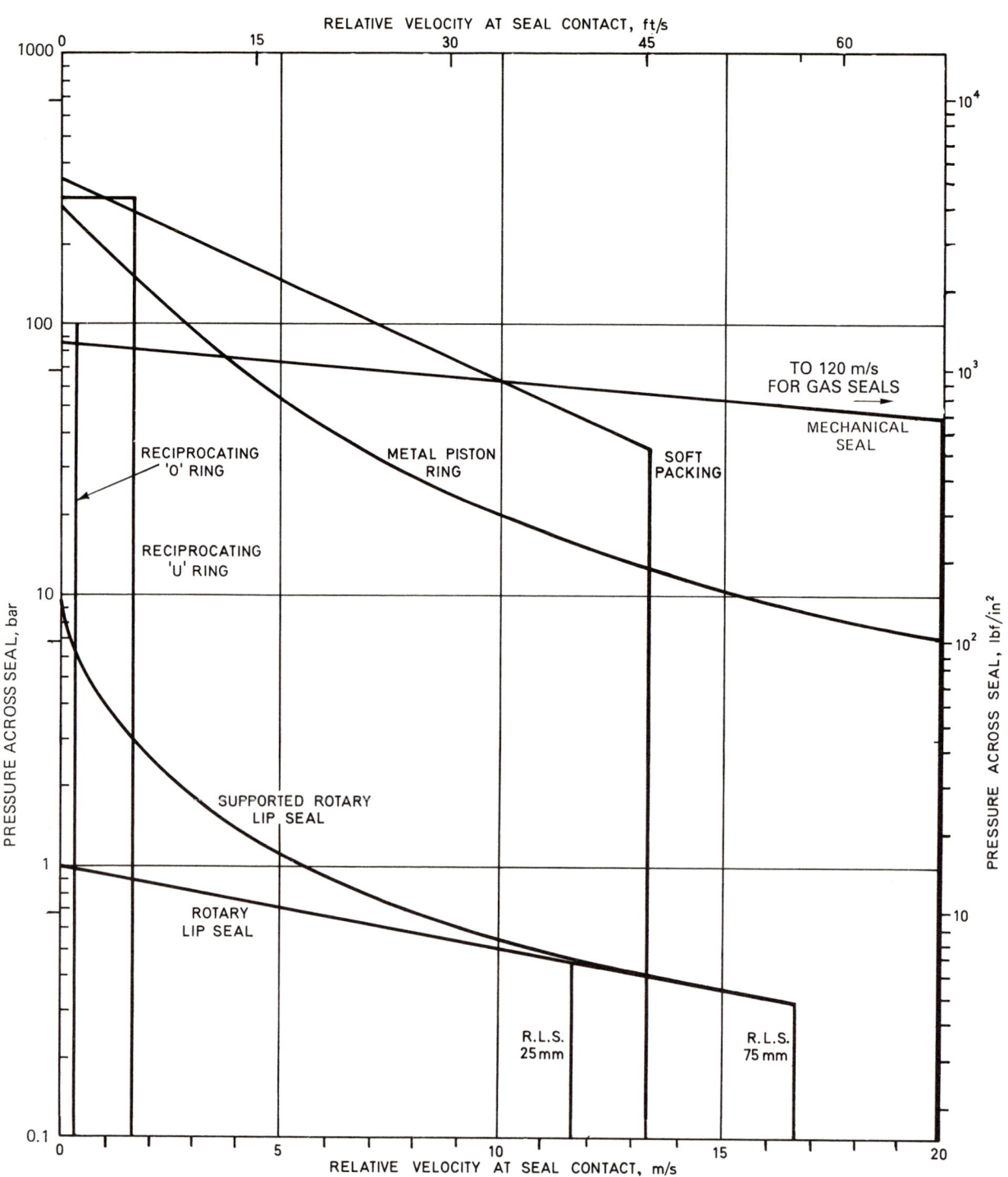

Figure 19.13 Limits of pressure and rubbing speed for various types of seal

Selection of seals 19

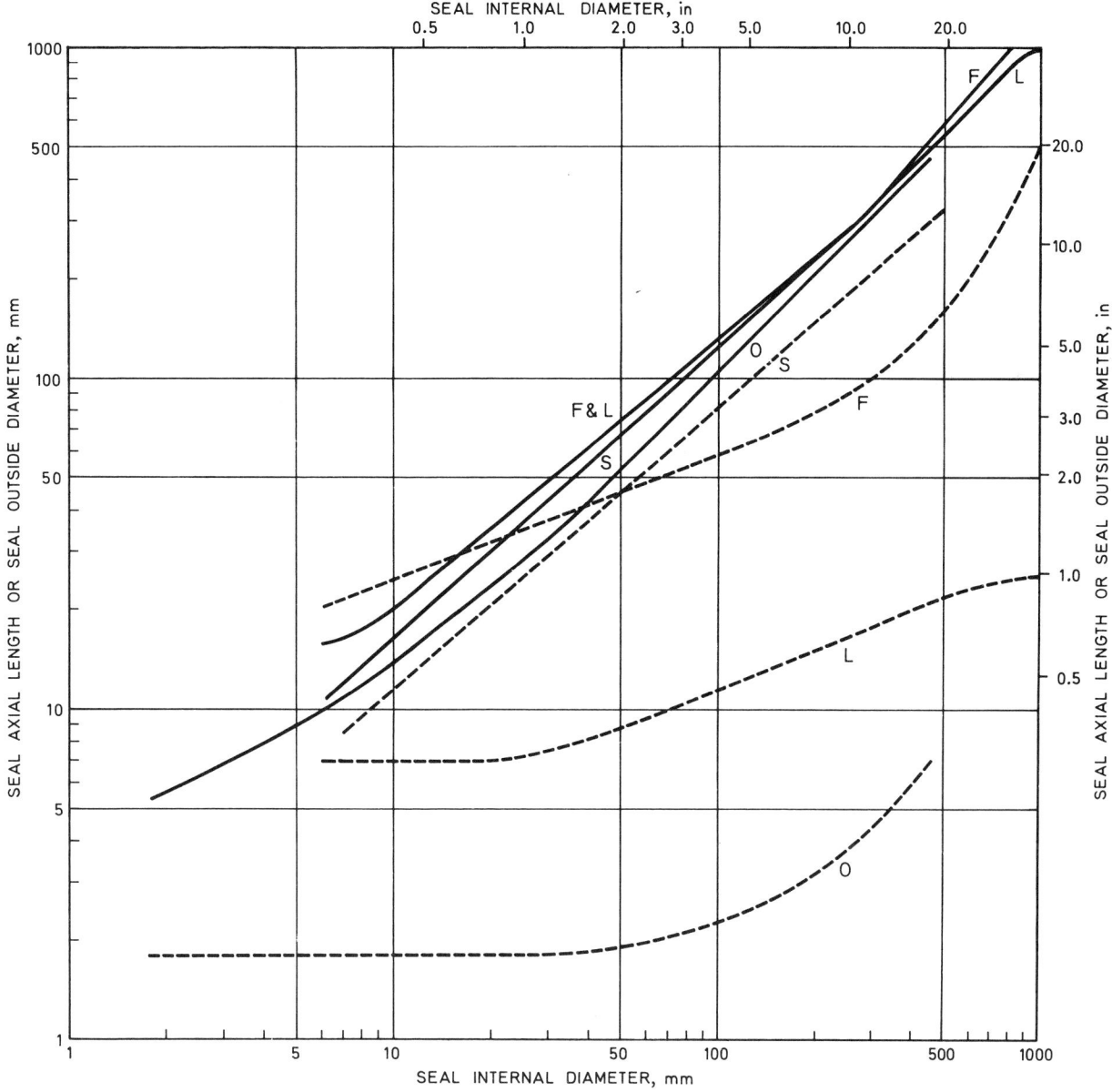

Figure 19.14 Normal minimum seal sizes (———, outside dia.; – – – –, length; F, mechanical seal; L, lip seal; O, 'O' ring; S, soft packing)

19 Selection of seals

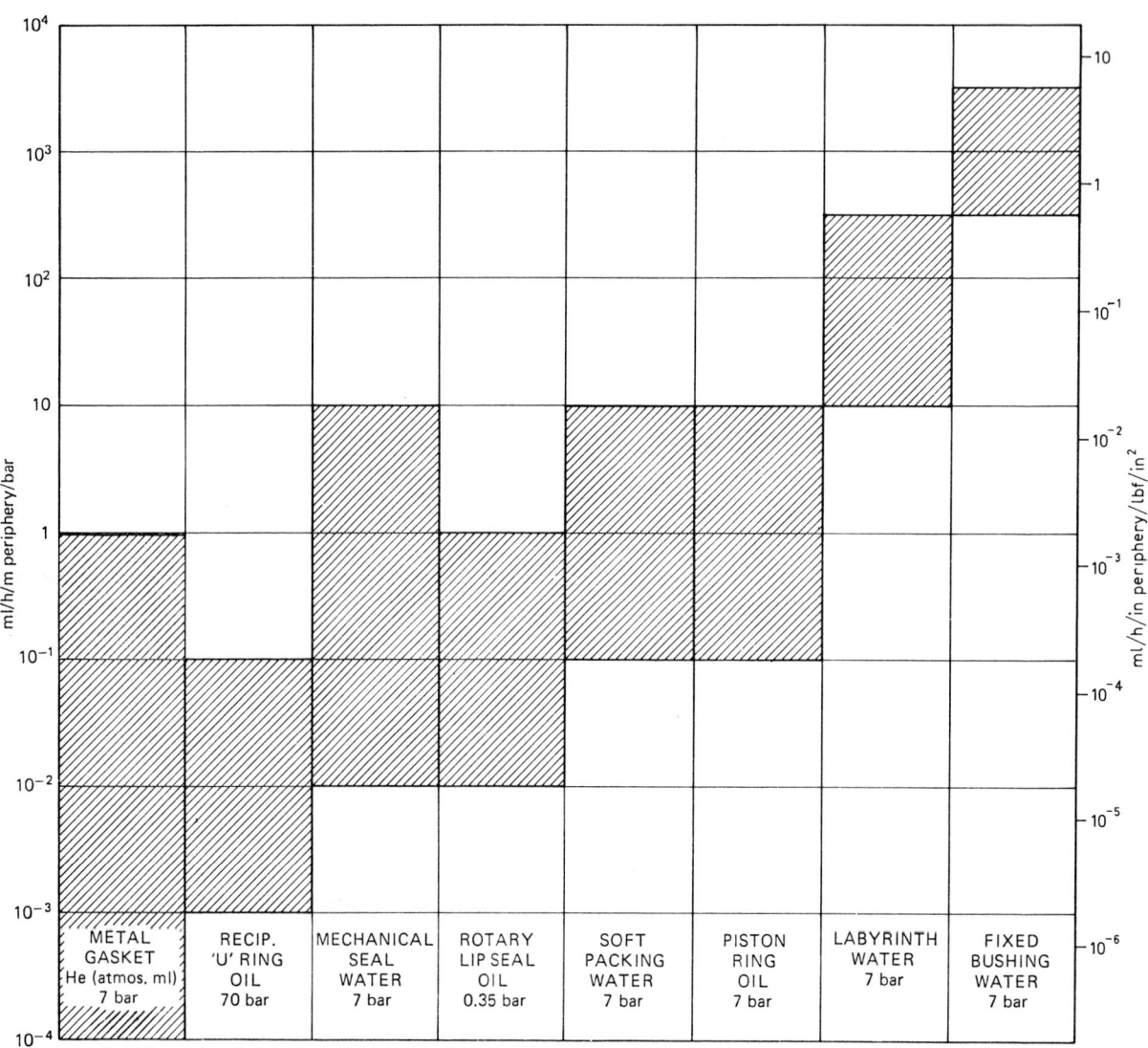

Figure 19.8 Approximate leakage rates for various types of seal

Sealing against dirt and dust   20

When operating in dirty and dusty conditions, the reliability of equipment depends almost entirely on the amount of abrasive material present. Natural soils contain abrasive materials in amounts varying from 98% down to 20% by weight.

Table 20.1 The source, nature and effect of contaminants

Source	Nature of contaminant	Operating conditions, effect on reliability and basic requirements		
		Wet (more than 15% by weight of water)	*Dry*	*Wet and dry*
Contact with soils (high silicon)	Sharp faceted grains predominantly silica (SiO_2) 98% by weight occurs frequently	Severe, highly abrasive. Considerable loss of reliability unless extensive sealing provided	Some loss of reliability. Machinery ingesting air requires efficient air cleaning	Severest, maximum abrasive effect. Very good air cleaning and sealing required
(low silicon)	Predominantly grains of Calcium (CaO). Silica less than 25% by weight.	Clogging rather than abrasive. Some loss of reliability. Sealing required	Little loss of reliability. Good air cleaners and sealing required	Poor conditions. Good air cleaning and sealing required
Airborne dust	Any finely divided material in the dry state picked up in air currents	—	Reduction in reliability dependent on dust concentration. Very large air cleaners required for highest concentrations	—

DESIGN OF SEALING SYSTEMS

Design to reduce the effects of dirt and dust

1. Keep to a minimum the number of rotary or sliding parts exposed to bad conditions.
2. Provide local clean environments for bearings and reciprocating hydraulic mechanisms by means of separate housings or sealing arrangements.
3. Provide adequate space in the sealing arrangement for oil lubrication.
4. Do not use grease lubrication for bearings, unless design for oil becomes uneconomic.
5. Provide adequate means for replenishment of lubricant; easily accessible.
6. Protect lubrication nipples locally to avoid erosion or fracture from stones and soil.
7. Provide positive means for checking amount of lubricant in housing.
8. Never use a common hydraulic fluid system for such mechanisms as reciprocating hydraulic motors and exposed hydraulic rams for earth moving equipment. Abrasive material is bound to enter the ram system which will be highly destructive to precision mechanisms. Provide independent fluid systems.
9. For mechanisms relatively crude in function where lubricant retention of any sort is either too costly or impracticable, load carrying bearings and reciprocating parts may be made in material with very hard or work hardening contact surfaces. Austenitic manganese steels have work hardening properties, but are not readily machinable. The shape of parts must be arranged to be used as cast or with ground surfaces.
10. Arrange the position of air cleaner intakes to avoid locally induced dust clouds from the motion of the mechanism.

20 Sealing against dirt and dust

Table 20.2 Sealing of rotary parts

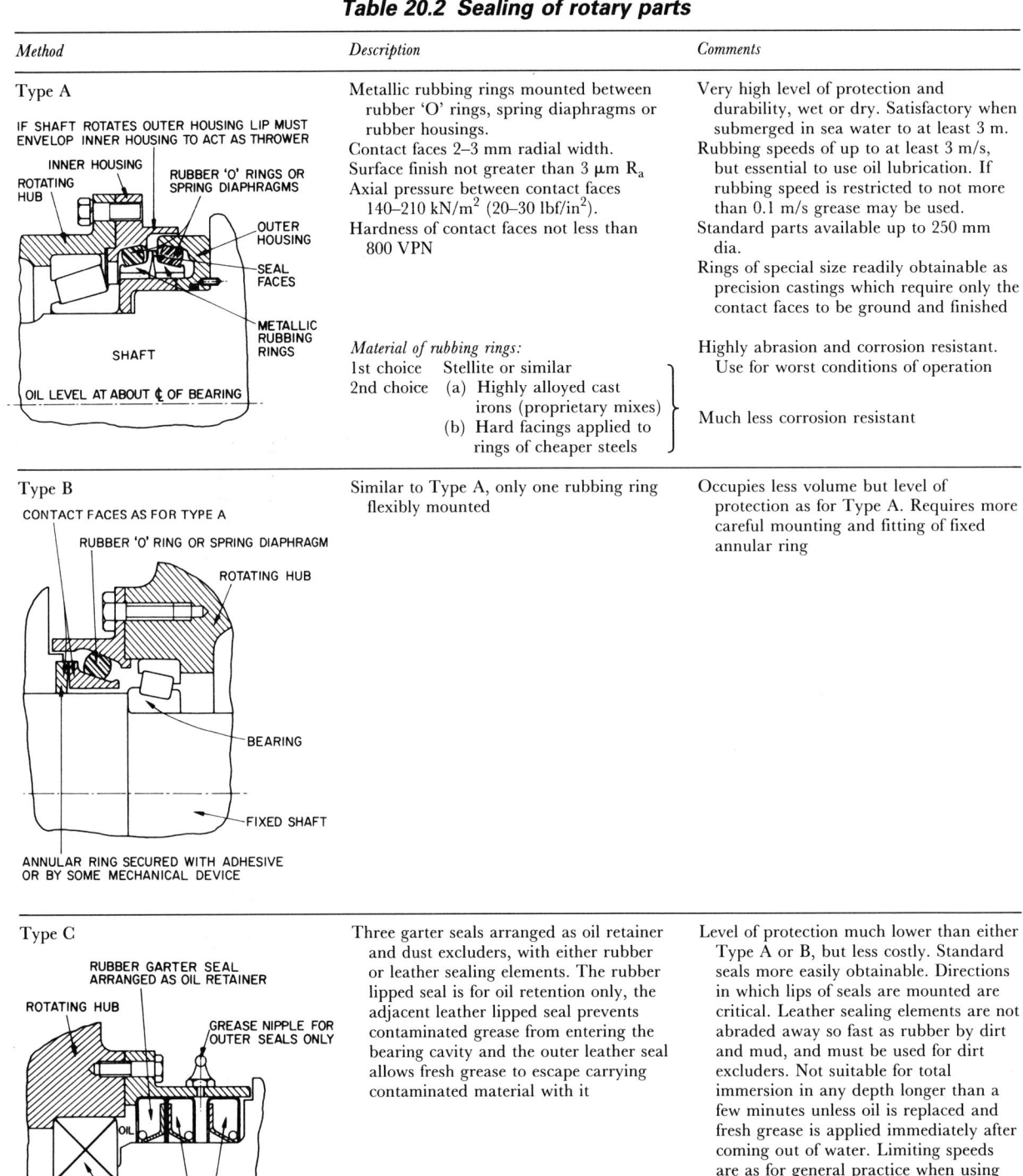

Method	Description	Comments
Type A	Metallic rubbing rings mounted between rubber 'O' rings, spring diaphragms or rubber housings. Contact faces 2–3 mm radial width. Surface finish not greater than 3 µm R_a Axial pressure between contact faces 140–210 kN/m^2 (20–30 lbf/in^2). Hardness of contact faces not less than 800 VPN Material of rubbing rings: 1st choice Stellite or similar 2nd choice (a) Highly alloyed cast irons (proprietary mixes) (b) Hard facings applied to rings of cheaper steels	Very high level of protection and durability, wet or dry. Satisfactory when submerged in sea water to at least 3 m. Rubbing speeds of up to at least 3 m/s, but essential to use oil lubrication. If rubbing speed is restricted to not more than 0.1 m/s grease may be used. Standard parts available up to 250 mm dia. Rings of special size readily obtainable as precision castings which require only the contact faces to be ground and finished Highly abrasion and corrosion resistant. Use for worst conditions of operation Much less corrosion resistant
Type B	Similar to Type A, only one rubbing ring flexibly mounted	Occupies less volume but level of protection as for Type A. Requires more careful mounting and fitting of fixed annular ring
Type C	Three garter seals arranged as oil retainer and dust excluders, with either rubber or leather sealing elements. The rubber lipped seal is for oil retention only, the adjacent leather lipped seal prevents contaminated grease from entering the bearing cavity and the outer leather seal allows fresh grease to escape carrying contaminated material with it	Level of protection much lower than either Type A or B, but less costly. Standard seals more easily obtainable. Directions in which lips of seals are mounted are critical. Leather sealing elements are not abraded away so fast as rubber by dirt and mud, and must be used for dirt excluders. Not suitable for total immersion in any depth longer than a few minutes unless oil is replaced and fresh grease is applied immediately after coming out of water. Limiting speeds are as for general practice when using seals of this type. In worst environment grease replenishment required daily. Pump in until grease is seen to exude from outer housing

Sealing against dirt and dust 20

Table 20.3 Sealing of reciprocating parts

Method	Description	Comments
A relay system as in Figure 20.1	The hydraulic device is built into a housing and a relay system converts the reciprocating into rotary motion. The rotating parts are sealed as shown in Table 20.2	High level of protection. The primary hydraulic seal functions in clean conditions
A flexible cover system as in Figure 20.2	A flexible cover is mounted over the main hydraulic seals and means provided for breathing clean air through piping from the inside of the cover to a clean zone or through an air cleaner. The cover material is highly oil resistant and preferably reinforced with fabric	At maximum speed of operation provide piping to limit pressure difference to $35\ kN/m^2$ ($5\ lbf/in^2$) between inside and outside of cover. The hydraulic system is dependent on the cover and although small pin-holes will not seriously reduce protection in dry conditions they will cause early failure if present when operating under water or in wet soil conditions
Standard chevron seals, 'O' rings etc.	Non metallic reciprocating seals used singly or in groups. All sliding parts through and adjacent to the seals highly corrosion resistant	Suitable where some loss of hydraulic fluid is not critical. An adequate reserve of hydraulic fluid must be provided to keep system full
Flexible metallic or non-metallic diaphragm	Hydraulic system sealed off completely	Very high level of reliability but restricted to small usable movements depending on diameter of diaphragm

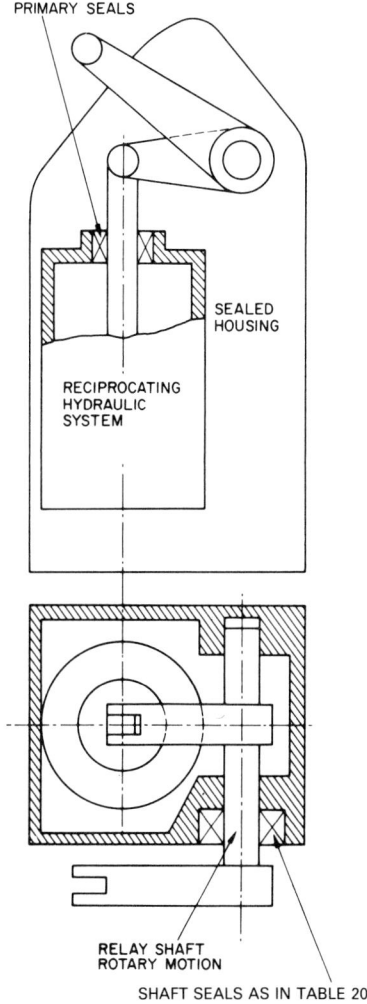

Figure 20.1 A relay system for reciprocating motion

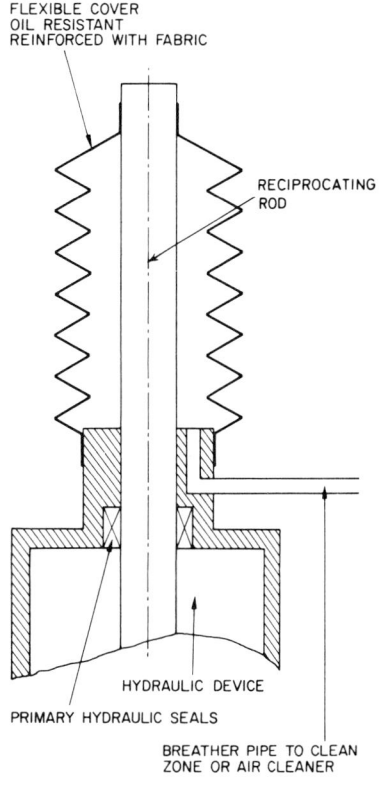

Figure 20.2 A vented flexible cover system

20 Sealing against dirt and dust

Table 20.4 Sealing with limited rotary or axial movement

Method	Description	Comments
Elastomeric deflection	Annular elastomer bushes either single or multi-layered, bonded or fastened to the adjacent parts.	Very high level of reliability in all environments at low cost. Elastomer must be matched to local contaminants. All motion either torsional or axial must occur in the elastomer. Usually, bushes made specially to suit load requirements

RECIPROCATING ENGINE BREATHING

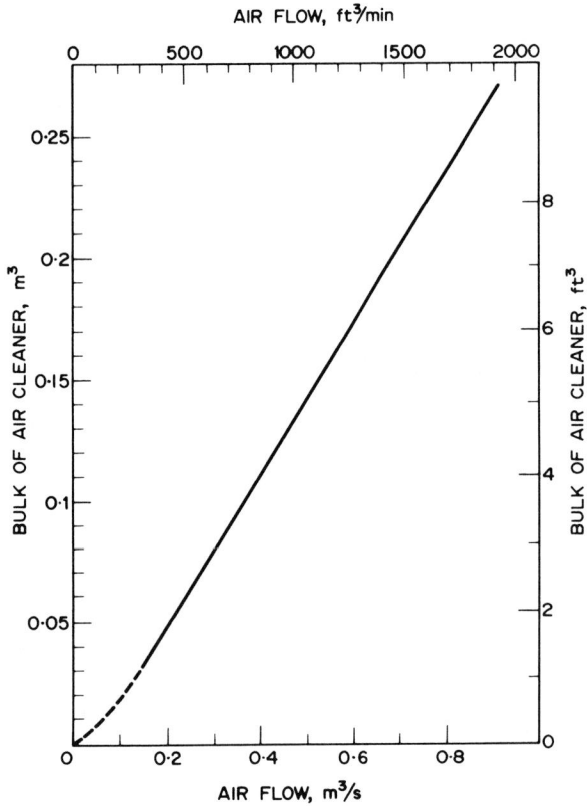

Dust concentration up to 0.0015 kg/m^3.

Type of cleaner: 2-stage, primary centrifuge with fabric secondary stage.

Fabric required for 2nd stage:
0.1 m^2 × 150 mm thick/37 kW.

Approx. relationship between air flow and bulk volume of complete cleaner shown in Figure 20.3.

Figure 20.3 Air cleaner requirements for reciprocating engine breathing; 20–100 h maintenance periods for max. dust concentration of 0.0015 kg/m^3, restriction 15–25 in w.g.

Oil flinger rings and drain grooves   21

Oil issuing from a bearing as end leakage will travel along a shaft for a finite distance before centrifugal dispersal of the film takes place. Many clearance seals will permit oil leakage from the bearing housing if they are situated within the shaft oil-film regime. Flinger rings and drain grooves can prevent the oil reaching the seal.

GENERAL PROPORTIONS

The natural dispersal length of the oil film along the shaft varies with the diameter and the speed as shown in Figure 21.1.

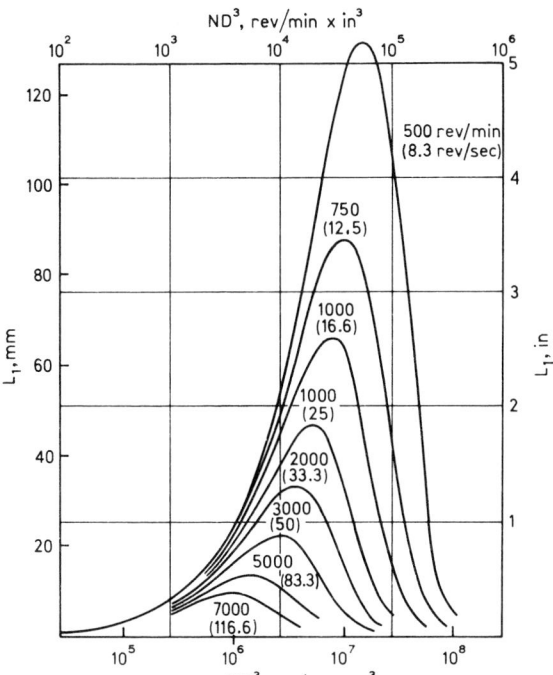

Figure 21.1

Notation:

L_1 = natural dispersal length of oil film—in (mm)
L_2 = distance of oil thrower from end of bearing—in (mm)
D = shaft diameter—in (mm)
D_0 = outside diameter of oil thrower—in (mm)
N = shaft speed—rev/min (rev/s)

Using the value of L_1 corresponding to the design value of ND^3 in Figure 21.1, the oil thrower diameter should be derived from:

$$D_0 = \sqrt{\frac{C}{N^2 D} \log_e \frac{L_1}{L_2} + D^2} \qquad (1)$$

where C has the value

        30×10^6 for inch rev/min units.
and 136×10^6 for millimetre, rev/s units.

In general, high-speed shafts require small throwers and low-speed shafts require large ones, particularly if the thrower is close to the bearing.

Where shafts must operate at any speed within a speed range, flingers should be designed by the foregoing methods using the minimum range speed.

Where shafts are further wetted by oil splash and where oil can drain down the inside walls of the bearing housing on to the thrower itself, larger thrower diameters than given by equation (1) are frequently employed. Figure 21.2 gives a guide to 'safe' thrower proportions to meet this condition.

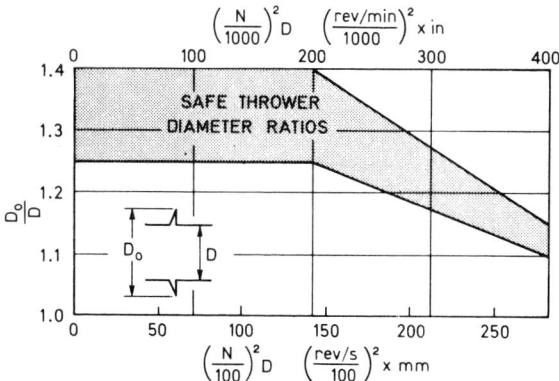

Figure 21.2

TYPICAL THROWERS

Scale details of some well-proven throwers are given in Figures 21.3, 21.4 and 21.5. Relevant values of D_0/D for the originals are given in each case. The application of each type may be assessed from Figure 21.2.

Type 1

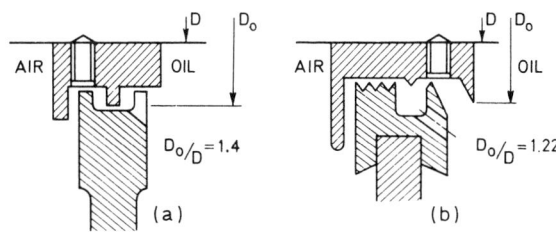

Figure 21.3 Throwers for slow/medium speeds

These are simple throwers of the slip-on type. Mild steel is the usual thrower material while a self-lubricating material such as leaded bronze is preferred for the split housing.

Note:

1. The drain groove from the annulus in (a) and the drain hole in (b).
2. The chamfer at the outer periphery of the (b) split housing to drain away oil washing down the walls of the bearing housing.
3. The chamfer at the back of the main thrower of (b) and the mating chamfer on the housing.

The above features are also common to the other types shown in Figures 21.4 and 21.5.

21 Oil flinger rings and drain grooves

Type 2

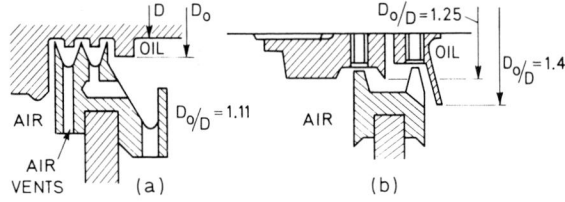

Figure 21.4 Throwers for medium/high speeds

Type 3

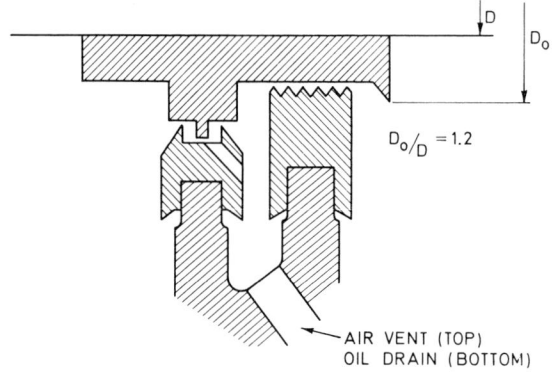

Figure 21.5 A medium/high speed two piece thrower

Note how the shaft enlargement on (a) has necessitated the introduction of a second annular space, vented to the atmosphere. Such enlargements, coupling hubs, etc. can create pressure depressions which can pull oil mist through the seal. Note the two-piece construction of (b) which gives a good sized secondary thrower. The shaped primary thrower is perhaps overlarge for a high-speed machine, but this is a good fault!

As an alternative to Type 2, a two-piece arrangement can be used if space permits. The primary seal can be of the visco seal or windback type. The secondary seal can be of the simple Type 1 variety. A substantial air vent is provided between the seals to combat partial vacuum on the air side.

DETAIL DIMENSIONS

Drain hole/oil groove sizing

Hole/groove area $\geq k \times$ thrower annular clearance area, corresponding to maximum design tolerances.

Suggested variation of k with shaft speed is given in Figure 21.6.

The individual diameter of the several drain holes making up the above area should not be less than 5 mm ($\frac{3}{16}$) in say.

Internal clearances

These are a matter of judgement. Suggested values for diametral clearance are:

Low-speed shafts: $1.25 \times$ max. design bearing diametral clearance.
High-speed shafts: $D/250$ or $2 \times$ max. design bearing diametral clearance, whichever is greater.

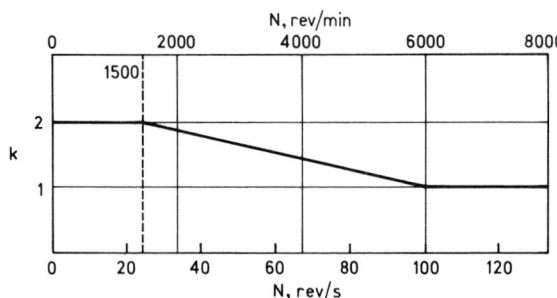

Figure 21.6

Labyrinths, brush seals and throttling bushes 22

PLAIN BUSH SEALS

Fixed bush seal

Leakage is limited by throttling the flow with a close-fitting bush—Figure 22.1.

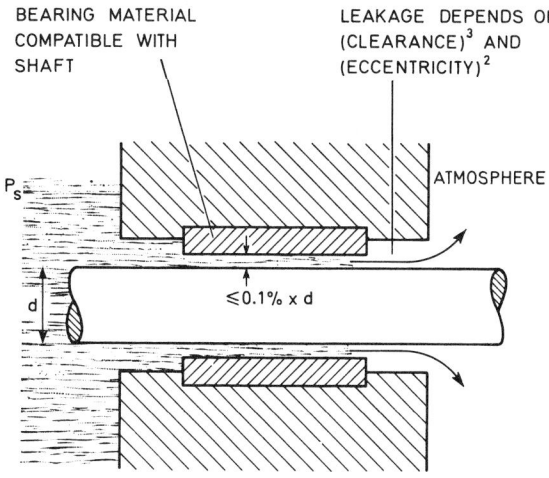

Figure 22.1 Typical fixed bush

Rigid floating bush seal

Alignment of a fixed bush can be difficult but by allowing some radial float this problem can be avoided (Figure 22.2)

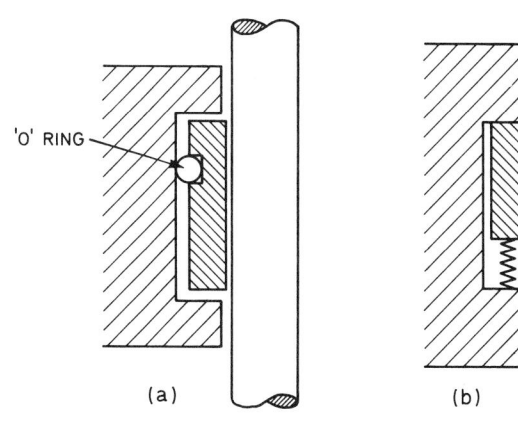

Figure 22.2 Bush seals with radial float

Leakage calculation

The appropriate formula is indicated in Table 22.1 for laminar flow conditions. For an axial bush with an incompressible fluid, Figure 22.3 can be used in both laminar and turbulent regions.

Table 22.1 Bush seal volumetric leakage with laminar flow

$q \equiv$ volumetric flow rate/unit pressure gradient/unit periphery $\eta \equiv$ absolute viscosity	*Fluid incompressible*	*Fluid compressible**
$\left(\epsilon = \dfrac{\delta}{c}\right)$ 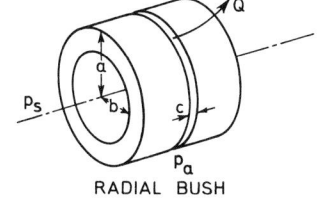 AXIAL BUSH	$Q = \dfrac{2\pi a(P_s - P_a)}{l} \cdot q$ m^3/s or in^3/s	
	† $q = \dfrac{c^3}{12\eta} \cdot (1 + 1.5\epsilon^2)$	$q = \dfrac{c^3}{24\eta} \cdot \dfrac{(P_s + P_a)}{P_a}$
RADIAL BUSH	$Q = \dfrac{2\pi a(P_s - P_a)}{(a - b)} \cdot q$ m^3/s or in^3/s	
	$q = \dfrac{c^3}{12\eta} \cdot \dfrac{(a - b)}{a \log_e \dfrac{a}{b}}$	$q = \dfrac{c^3}{24\eta} \cdot \dfrac{(a - b)}{a} \cdot \dfrac{(P_s + P_a)}{P_a}$

* For Mach number < 1.0, i.e. fluid velocity < local velocity of sound.

† If shaft rotates, onset of Taylor vortices limits validity to $\dfrac{V_c}{\nu}\sqrt{\dfrac{c}{a}} < 41.3$ (where ν = kinematic viscosity).

111

22 Labyrinths, brush seals and throttling bushes

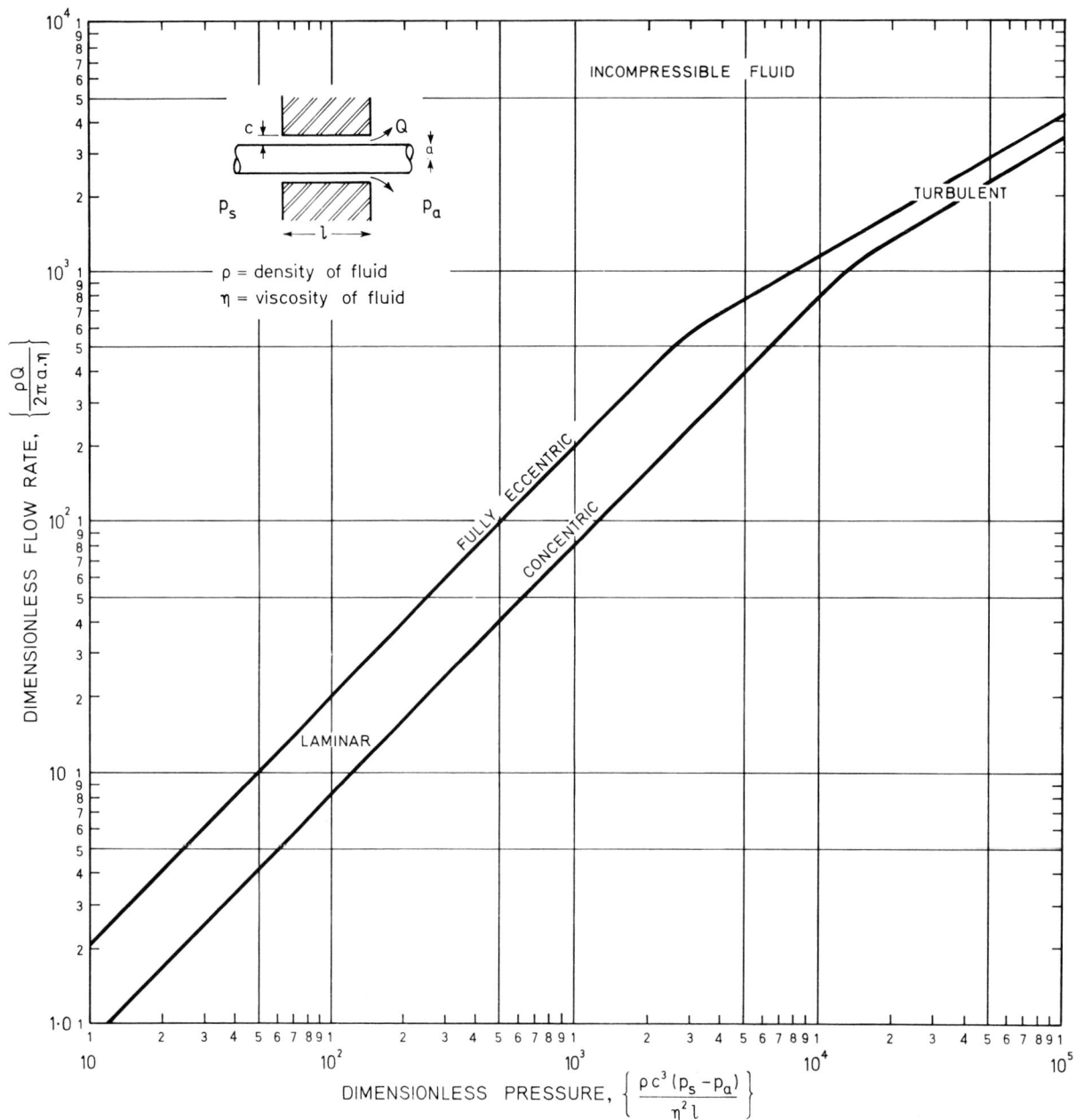

Figure 22.3 Leakage flow of an incompressible fluid through an axial bush at various shaft eccentricities

Labyrinths, brush seals and throttling bushes 22

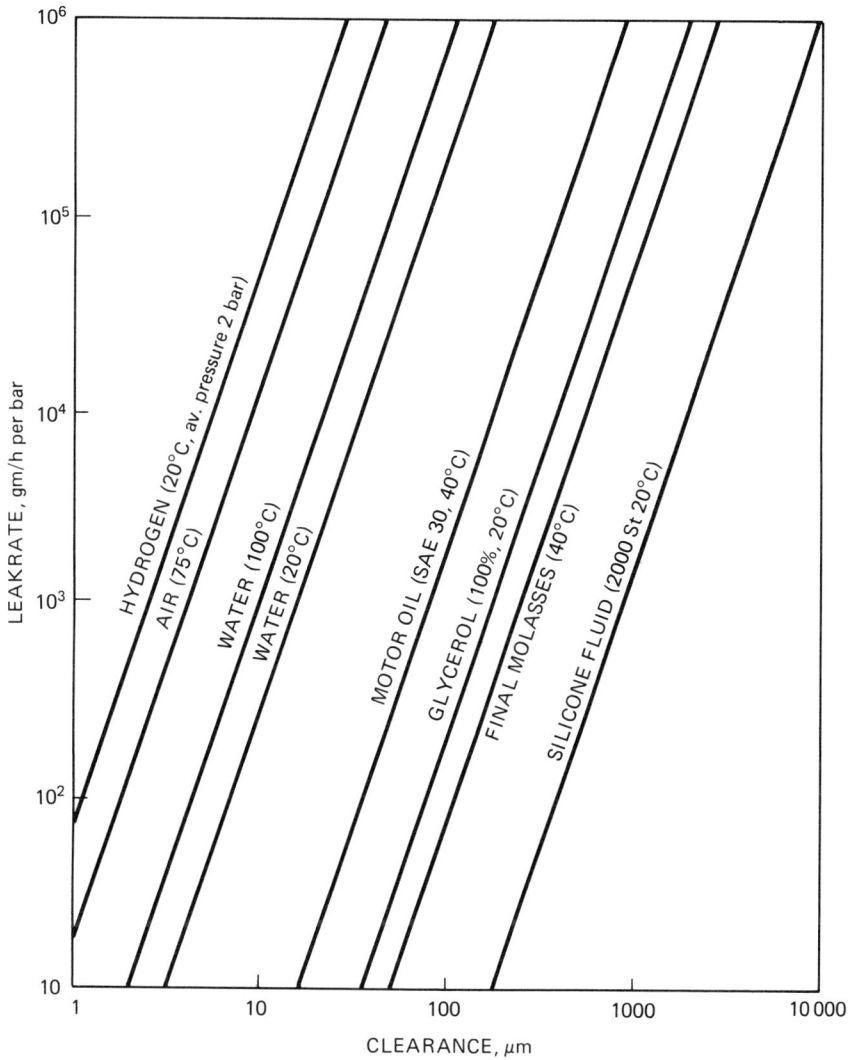

Figure 22.4 Effect of clearance and viscosity on mass flow leakage from a bush seal with length equal to perimeter (if perimeter = k × length then multiply leakrate by k)

Mass flow rate (M) through the clearance of a bush seal:

$$M = 9.4 \times 10^{-8} \frac{\rho h^3 \Delta_p D}{\eta L} \text{ gm/h}$$

where ρ = density, kg/m^3
 h = clearance, μm
 Δ_p = pressure differential, bar
 D = mean diameter, mm
 η = dynamic viscosity, $\dfrac{\text{Ns}}{\text{m}^2}$
 L = leakage path length, mm

Materials

Bush and shaft may sometimes rub; therefore compatible bearing materials should be selected. Segmented bushes are commonly made of carbon–graphite and may run on a shaft sleeve of centrifugally cast bronze (90% Cu, 10% Sn) when sealing water. The life of such a combination could be 5–10 years if abrasives are not present. Other suitable materials for the sleeve or shaft include carbon steels, for more severe conditions nitriding or flame hardening is recommended, and stainless steel. In the latter case a compatible babbit may be used to line the bush.

22 Labyrinths, brush seals and throttling bushes

BRUSH SEALS

Brush seals are an alternative for labyrinths in gas turbine engine applications, reducing leakage by a factor up to five or tenfold, although relatively expensive. The brush seal comprises a bundle of metal filaments welded at the base. The filaments are angled circumferentially at about 45 degrees, filament length is chosen to give an interference of 0.1–0.2 mm with the sealing counterface. Filaments are typically about 0.7 mm diameter and manufactured from such materials as high temperature alloys of nickel or cobalt. Suitable counterface materials include hardfacings of chromium carbide, tungsten carbide or alumina

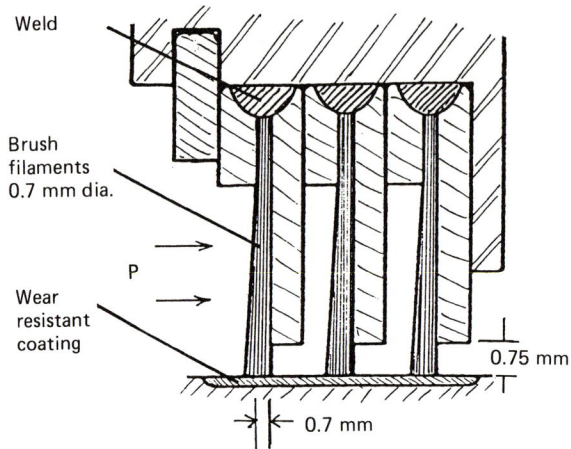

Figure 22.5 Brush seal with typical dimensions

LABYRINTHS

A labyrinth [Figure 22.6] can reduce leakage below that of a bush seal by about half. This is because eddies formed in the grooves between the vanes increase the flow resistance. Labyrinths are commonly used for gas and steam turbines.

Typically, the vane axial spacing might be twenty times the vane lip clearance. If the spacing is too small eddy formation is inefficient, if too large then the seal becomes unduly long.

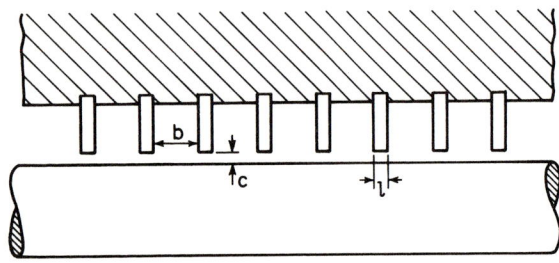

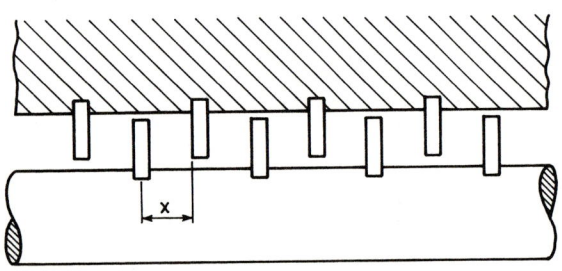

Figure 22.6 Typical labyrinth arrangements

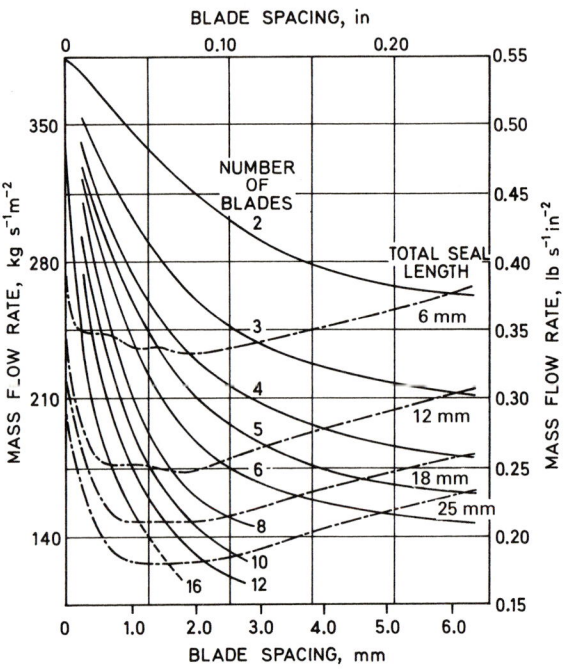

Ambient pressure = 1 bar; blade thickness = 0.14 mm
radial clearance = 0.127 mm   pressure ratio = 0.551;
temperature 310 K

Figure 22.7 Performance of a typical labyrinth seal

Materials

Vane and rotor must be compatible bearing materials, in case of rubbing contact, but in turbines stresses are high and creep is a critical factor. Metal foil honeycomb is a convenient form of material for labyrinths since its integrity is retained if rubbing occurs, and yet it deforms readily. Suitable honeycomb is produced in 18/8 stainless steel and Nimonic 75, 80a and 90. Another convenient combination comprises metal fins and a carbon bush. The latter can be segmented if necessary for large diameters.

Labyrinths, brush seals and throttling bushes 22

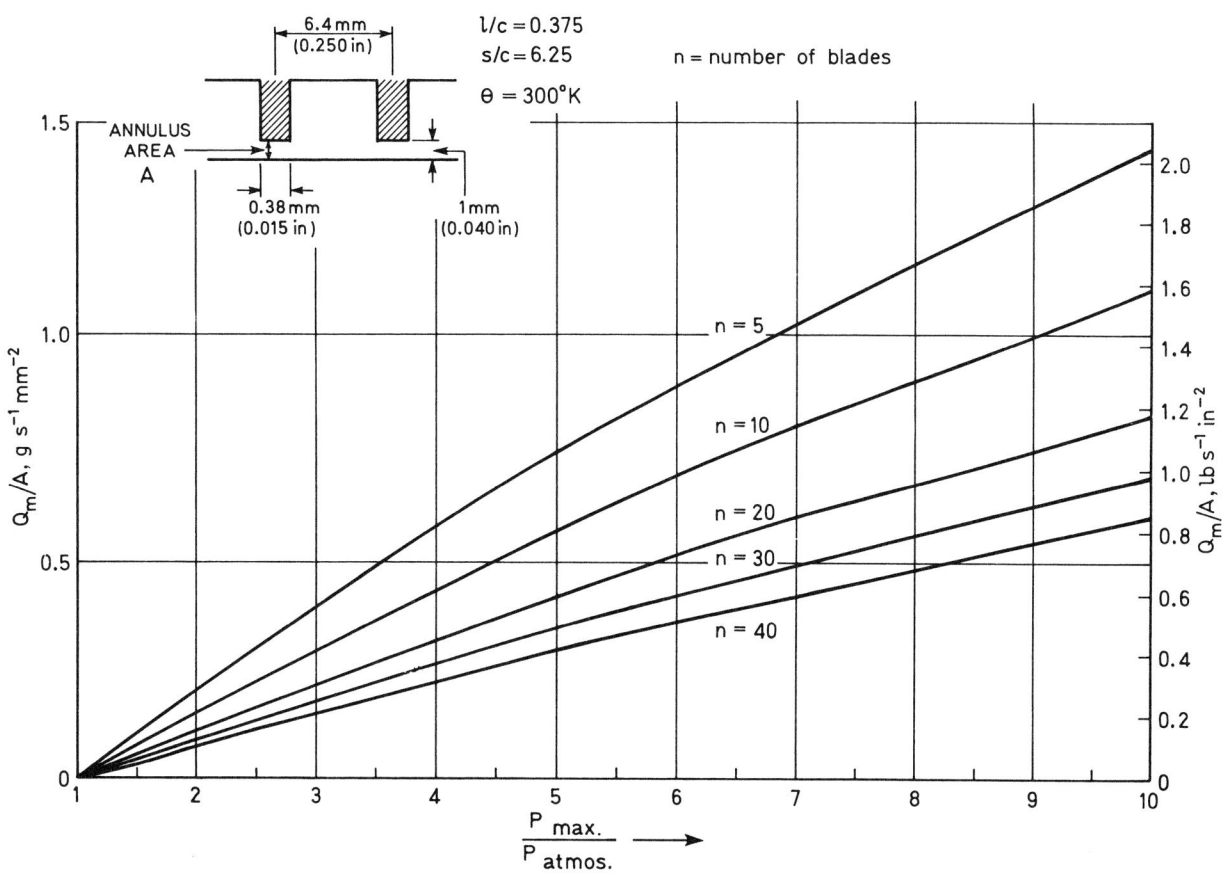

TEMPERATURE

For other temperatures multiply Q_m/A by $\dfrac{17.3}{\sqrt{\theta_k}}$

GEOMETRY

For other values of s/c or l/c multiply Q_m/A by g:
(see below left)

STEPPED LABYRINTH

For a stepped labyrinth use an 'effective' n value, kn:
(see below right)

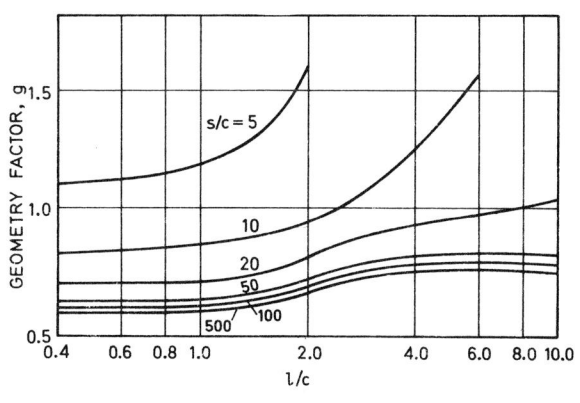
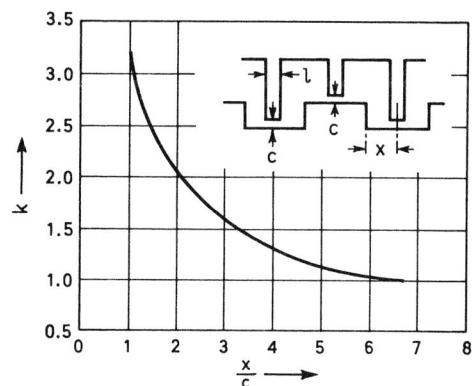

Figure 22.8 Calculation of mass flow rate of labyrinth

22 Labyrinths, brush seals and throttling bushes

VISCOSEAL
(Also called a screw seal or wind back seal)

Resembles a bush seal in which a helical groove has been cut in the bore of the bush or on the shaft (Figure 22.9). As the shaft rotates the helix pumps any leaking fluid back into the sealed system. There is no *sealing action* if the shaft is not rotating: an auxiliary seal can be fitted to prevent static leakage and may be arranged to lift off automatically when the shaft rotates, thereby reducing wear.

Viscoseals are used with *viscous fluids*, or at *high rotational speeds*, to seal low or moderate presures. When the pumping action just balances the leakage flow with the helix full of liquid there is no net leakage—this is the *sealing pressure*. If the system pressure exceeds the sealing pressure the seal leaks. At lower pressures the helix runs partially dry.

Optimum design and performance prediction

Helix angle, β = 15.7°
Clearance ratio, h/c = 3.7
Land ratio b/a = 0.5
Groove profile rectangular

The 'sealing pressure' for a seal of these dimensions, is given by:

$$p_s = \frac{6\eta V L}{c^2} \cdot k \quad (\eta = \text{absolute viscosity})$$

(i) $k = 0.1$ if $(Vc\rho/\eta) < 500$ (i.e. laminar flow)
or (ii) k is given by Figure 22.10 if $(Vc\rho/\eta) > 500$ (i.e. turbulent flow).

The optimum design varies with Reynolds number $[(Re) = Vc\rho/\eta]$ for turbulent flow; in particular a smaller helix angle than the optimum laminar value is better.

Eccentricity will lower the sealing pressure with laminar flow but may raise it for turbulent flow.

A problem may arise due to 'gas ingestion' when running in the sealing condition. This is due to instability at the liquid/air interface and is not easily prevented. The effect is to cause air to be pumped into the sealed system.

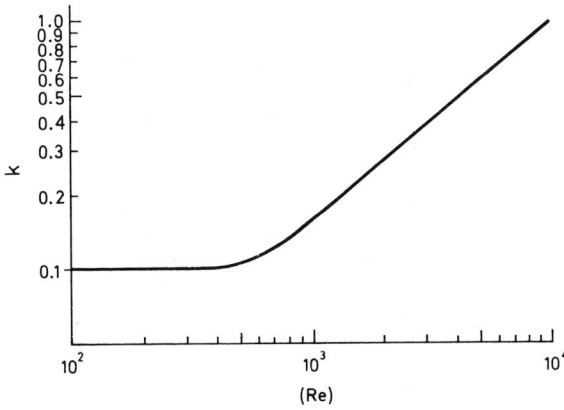

Figure 22.10 Variation of k with Reynolds Number

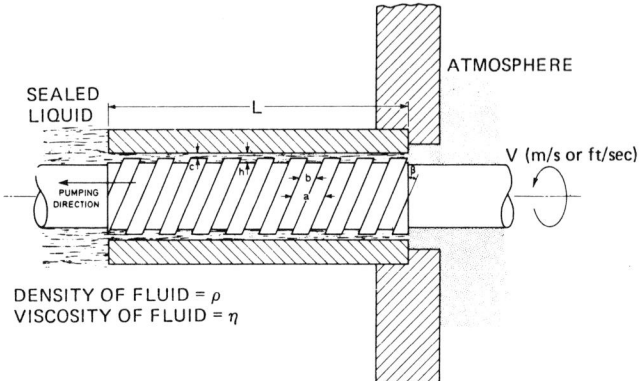

Figure 22.9 Viscoseal or wind-back seal

Typical performance

A 25 mm dia. seal with a clearance of 0.025 mm, length 25 mm, sealing water with a shaft speed of 2.5 m/s would be expected to seal at pressures up to 0.67 atm.

BARRIER VISCOSEAL

By installing a pair of conventional viscoseals back to back, so that pressure is built up between them, a pressure barrier forms to prevent the sealed fluid escaping (Figure 22.11). The sealed fluid must not be miscible with the barrier fluid. Typically the former is a gas and the latter a liquid or grease.

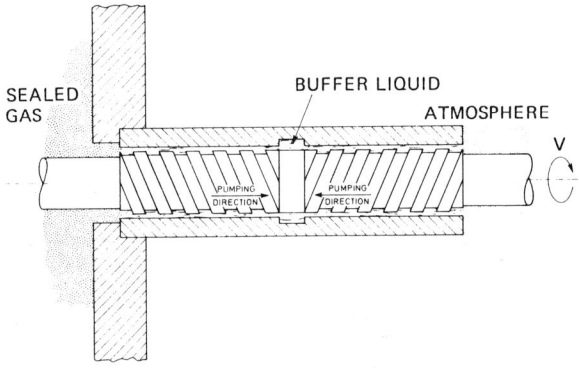

Figure 22.11 Barrier viscoseal

Typical performance

In tests using grease as the barrier fluid, gas pressures up to 10 atm have been sealed by a 13 mm dia. seal at 1000 rev/min.

Lip seals 23

SEALS FOR ROTATING SHAFTS

Design variations

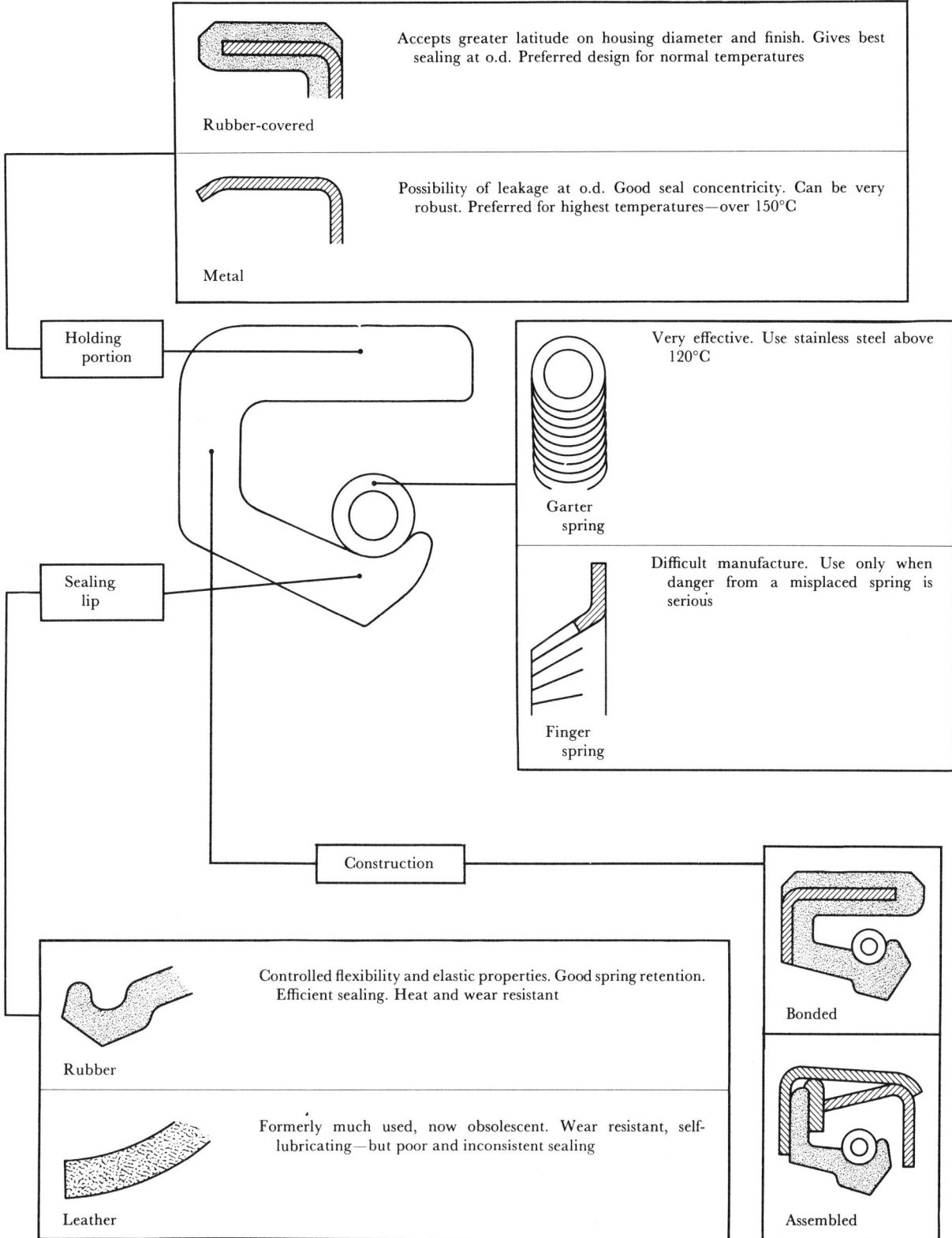

The above variants are the important ones to the seal user or machine designer, apart from special seals for dirty conditions. There are multitudinous detail modifications introduced by seal manufacturers, but for normal applications these can be ignored.

23 Lip seals

Operating conditions

 Max. speed	Up to 35 mm ($1\frac{1}{4}$ in) dia. 75 mm (3 in) dia. Over 75 mm (3 in) dia.	Approx. 8000 rev/min Approx. 4000 rev/min Approx. 15 m/s (50 ft/s) peripheral speed
 Max. fluid pressure	Up to 75 mm (3 in) dia. Over 75 mm (3 in) dia.	Approx. 0.6 bar (10 p.s.i.) Approx. 0.3 bar (5 p.s.i.) By using a profiled backing washer to support the lip, pressures up to 6 bar (100 p.s.i.) can be accommodated
 Temperature range	See table of rubber materials	Permissible oil temperatures are set by the sealing lip material. Do not ignore low-temperature conditions
 Eccentricity	Housing	Better than 0.25 mm (0.010 in) total indicator reading when clocked from shaft
	Shaft	Depends on speed. Aim for better than 0.025 mm (0.001 in) total indicator reading when rotated in its own bearings
Surface finish	Housing	Fine turned. Provide lead-in chamber
	Shaft	Grind and polish to better than 0.5 μm R_a. Surface must be free from all defects greater than 0.0025 mm (0.0001 in) deep. Use cardboard protection sleeve during manufacture
Machining tolerances	Housing	Up to 100 mm (4 in) ±0.025 mm (0.001 in) 100–175 mm (4–7 in) ±0.037 mm (0.0015 in) Over 175 mm (7 in) ±0.05 mm (0.002 in)
	Shaft	Up to 50 mm (2 in) ±0.025 mm (0.001 in) 50–100 mm (2–4 in) ±0.037 mm (0.0015 in) 100–200 mm (4–8 in) ±0.05 mm (0.002 in) Over 200 mm (8 in) ±0.125 mm (0.005 in)

Sealing dirt and grit

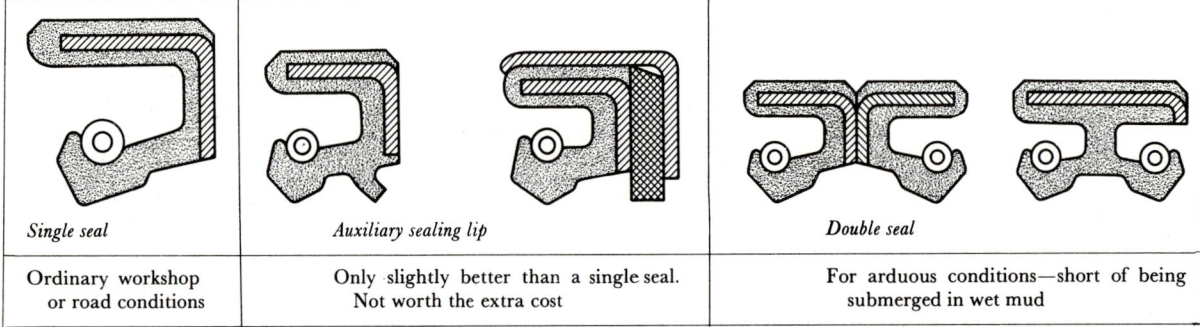

Single seal	*Auxiliary sealing lip*	*Double seal*
Ordinary workshop or road conditions	Only slightly better than a single seal. Not worth the extra cost	For arduous conditions—short of being submerged in wet mud

Lip seals 23

Sealing lip material (rubber)

Type of rubber	Typical trade names	Working temperature range, °C	Resistance to: Mineral oil	Resistance to: Chemical fluids	Relative cost of seal
Nitrile	Hycar Polysar	−40 to +100	Excellent	Fair	1
Acrylate	Krynac Cyanacryl	−20 to +130	Excellent	Fair	2
Fluoropolymer	Viton	−30 to +200	Excellent	Excellent	10
Polysiloxane	Silastomer	−70 to +200	Fair	Poor	4

Nitrile synthetic rubber is the universal choice for sealing oil or grease at temperatures below 100°C. For more extreme temperature conditions the choice is normally one of the other materials shown in the table—with some penalty in other directions. When in doubt consult the seal manufacturer.

Positive-action seals

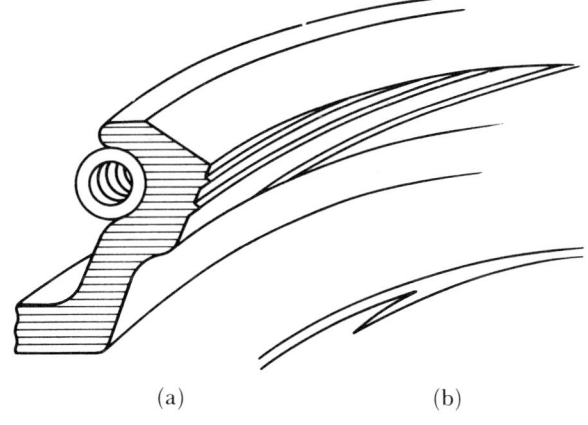

(a) (b)

Figure 23.1
(a) Section of a positive-action seal, showing the helical ridges on the air side.

(b) View of contact band through a glass shaft, showing one of the thread run-outs.

The 'positive-action' feature improves sealing performance and reliability. It is essential in conditions where eccentricity or vibration is beyond the limit for a normal seal.

Storage and fitting

1. Store in a cool place in manufacturer's package.
2. Lubricate before installing. Handle carefully.
3. Use a sleeve on the shaft to protect seal from damage by sharp edges, keyways, etc.
4. Press home squarely into housing, using a proper tool.

23 Lip seals

SEALS FOR RECIPROCATING SHAFTS
Packing types

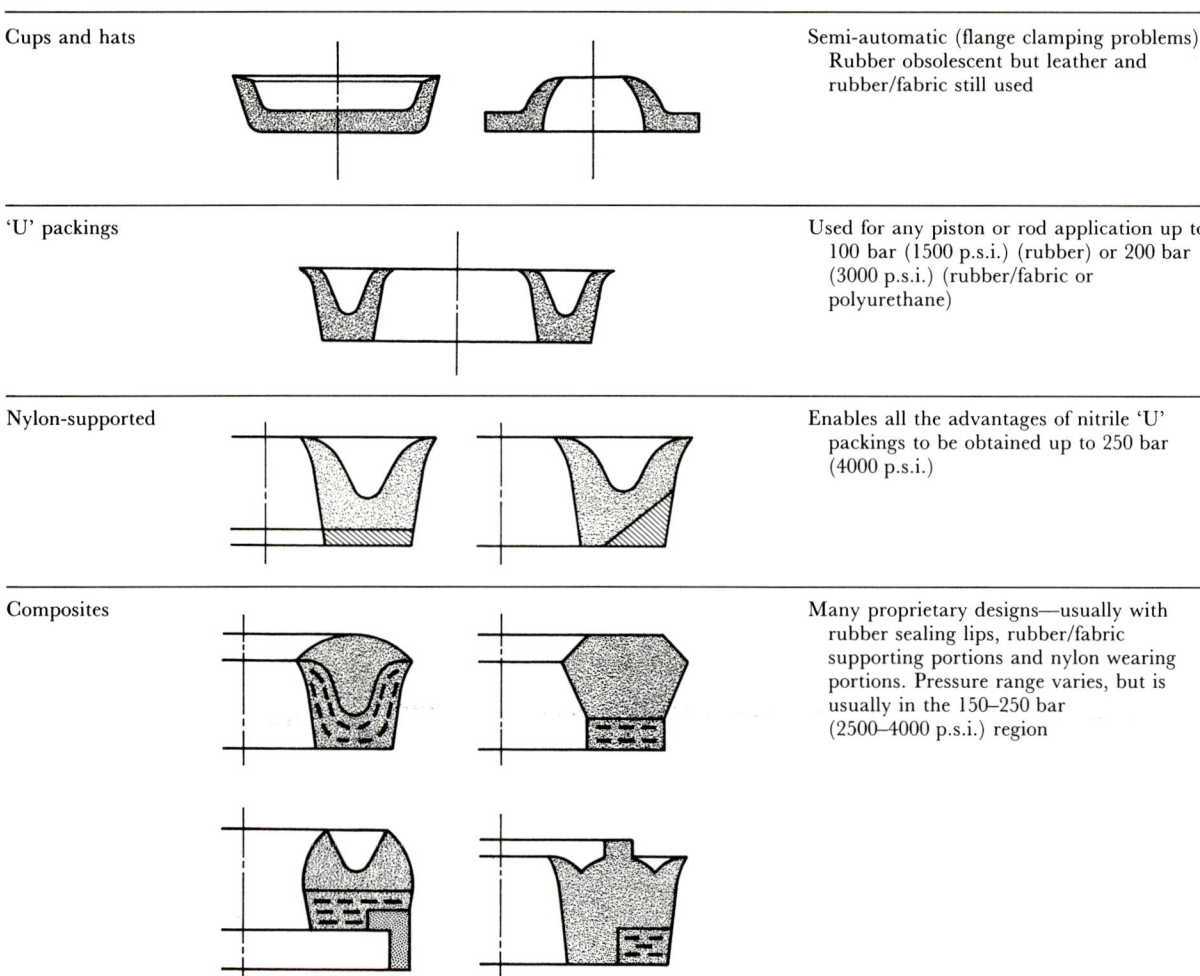

Cups and hats		Semi-automatic (flange clamping problems). Rubber obsolescent but leather and rubber/fabric still used
'U' packings		Used for any piston or rod application up to 100 bar (1500 p.s.i.) (rubber) or 200 bar (3000 p.s.i.) (rubber/fabric or polyurethane)
Nylon-supported		Enables all the advantages of nitrile 'U' packings to be obtained up to 250 bar (4000 p.s.i.)
Composites		Many proprietary designs—usually with rubber sealing lips, rubber/fabric supporting portions and nylon wearing portions. Pressure range varies, but is usually in the 150–250 bar (2500–4000 p.s.i.) region

Materials

Rubber (nitrile)	Highest sealing efficiency. Easily formed to shape. Low cost. Limited pressure capability, 100 bar (1500 p.s.i.). Excellent wear resistance. Poor extrusion resistance
Rubber-impregnated fabric	Great toughness, resistance to cutting and extrusion. Wear resistance inferior to plain rubber
Polyurethane rubber	Best toughness and wear resistance. Used for flexible packings at the highest pressures
Leather	Good wear and extrusion resistance. Limited shaping capability. Poor resistance to permanent set
Nylon	Combines well with rubber to resist extrusion and to provide good bearing surfaces

Lip seals 23

Extrusion clearance—mm(in)

	Up to 100 bar (1500 p.s.i.)		100–200 bar (1500–3000 p.s.i.)		Over 200 bar (3000 p.s.i.)	
	Normal	*Short life*	*Normal*	*Short life*	*Normal*	*Short life*
Rubber	0.25 (0.010)	0.5 (0.020)	—	—	—	—
Rubber/fabric leather	0.4 (0.015)	0.6 (0.025)	0.25 (0.010)	0.5 (0.020)	0.1 (0.005)	0.25 (0.010)
Polyurethane	0.4 (0.015)	0.6 (0.025)	0.25 (0.010)	0.5 (0.020)	0.1 (0.005)	0.25 (0.010)
Nylon support	—	—	0.25 (0.010)	1.0 (0.040)	0.1 (0.005)	0.5 (0.020)

Design of metal parts

	Cylinders	*Piston rods*
Preferred materials	Steel Cast iron	Steel
Heat treatment	Not required	Harden if possible
Plating	Not required	Chrome plate if possible
Surface finish	Grind or hone 0.5 μm R_a max.	Grind and polish 0.5 μm R_a max.
Machining tolerances	Fixed by extrusion clearance—see previous table	

Friction

Friction varies considerably with working conditions but for preliminary design purposes it can be assumed that it will be between 0.5% and 3% of the load which would be produced on a piston of the same diameter by the fluid pressure involved. For more accurate values the seal manufacturer must be consulted.

Points the designer should watch

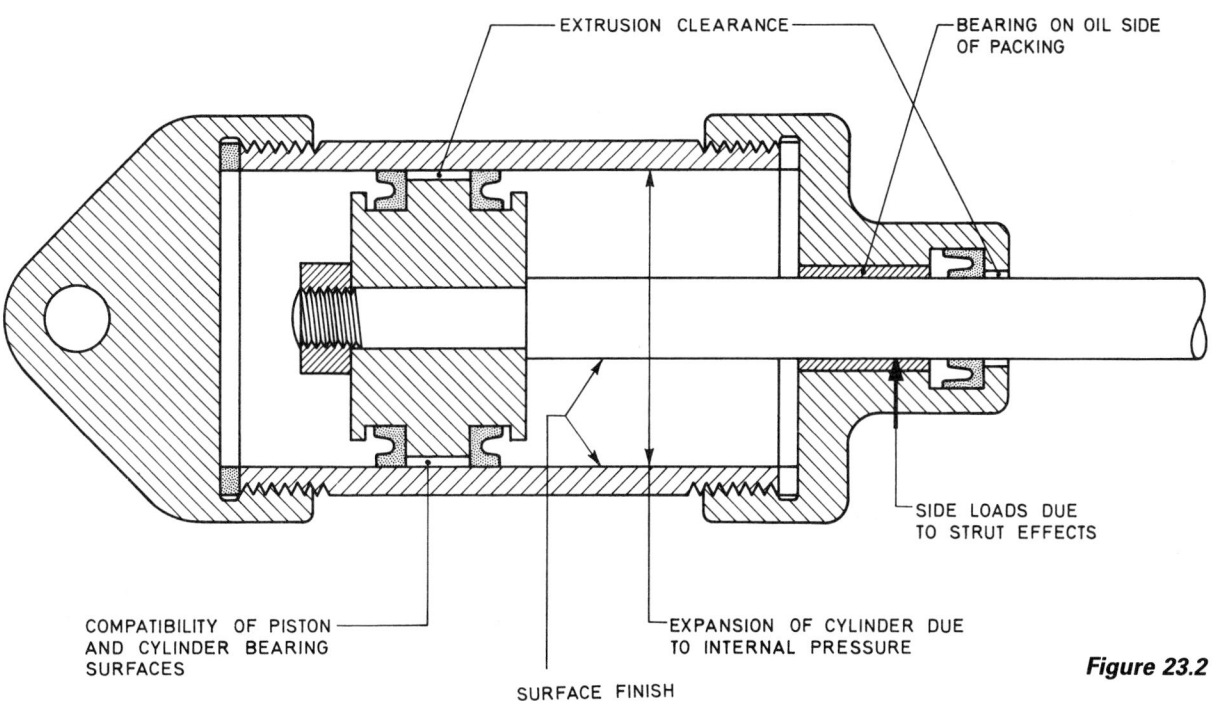

Figure 23.2

24 Mechanical seals

TYPES OF MECHANICAL SEAL

Table 24.1 Standard types of mechanical seal

Type of seal	Maximum allowable sealed pressure	Remarks
UNBALANCED SEALS	10–14 bar 140–200 lbf/in^2	Design is more simple than a balanced seal. Can be fitted directly to a constant diameter shaft. Rubber bellows type is less sensitive to build up of contaminants on the shaft surface
BALANCED SEALS	70 bar 1000 lbf/in^2	Balanced design reduces load on seal faces and enables higher pressures to be sealed. Design is more complex and shaft sleeve is essential
BACK TO BACK SEAL	As above	If fluid to be sealed is: a gas, a liquid near its boiling point, abrasive, a poor lubricant, this arrangement enables a separate liquid sealant (such as oil) to be used. It is, however, rather complex. Environmental awareness and safety requirements make the use of this seal more necessary.

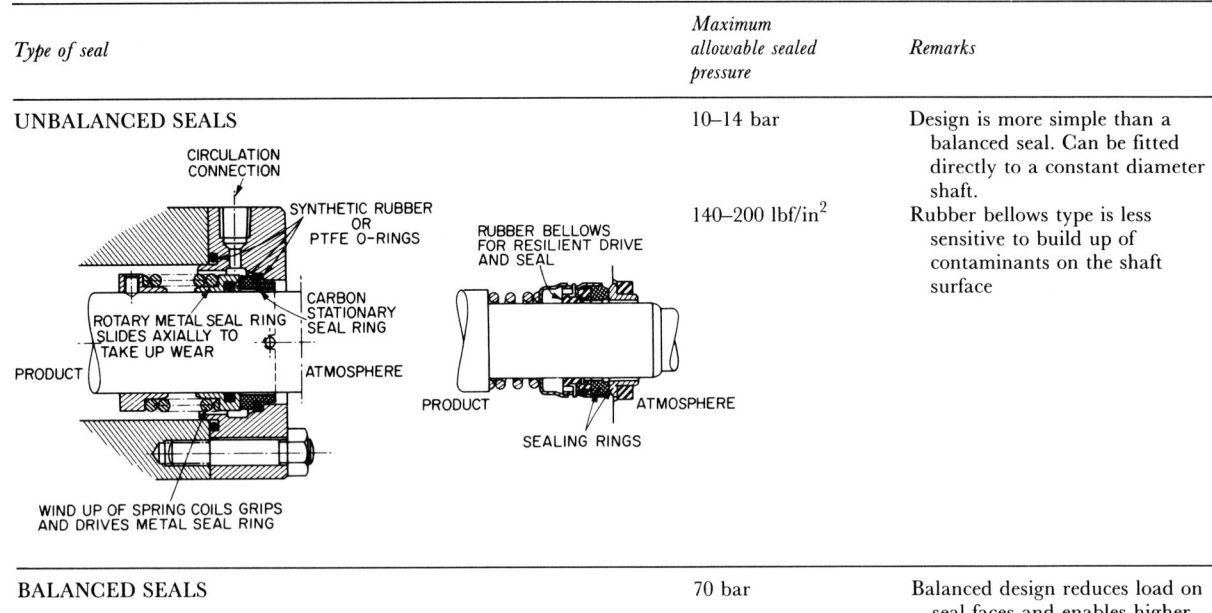

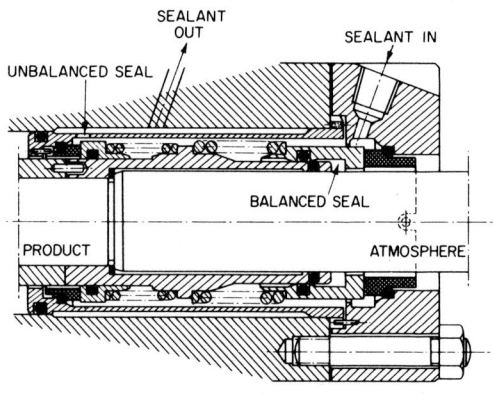

Mechanical seals 24

Table 24.2 Special seals and additional design features

Feature	Reason for inclusion	Remarks
CERAMIC/PTFE SEAL	Excellent corrosion resistance of components in contact with the fluid to be sealed	Particularly suitable for chemical applications
SPLIT SEAL	Ease of replacement of wearing components	The splitting of both sealing components reduces allowable performance and gives some leakage
FLUSH	To flush sediments and other debris from the sealing zone and to remove frictional heat. To keep seals handling hot products cool. To keep seals handling cold products warm	On pumps with the seal on suction pressure, a small flow of product from pump discharge is piped to the seal, and returns to pump suction through the balance holes in the impeller. If the differential head exceeds 15 m a flow controller is needed to limit the flow rate. This is only suitable if the product is free of particles above 50 μm. With dirty products a cyclone separator should be used instead, to provide clean product to the seal and return dirty product to the suction
QUENCH	To remove or prevent the build up of crystals or decomposition products on the atmospheric side of the seal	Water flushing will generally remove crystal deposits and low pressure steam will remove 'coke' from hydrocarbons
SAFETY BUSH (illustrated on diagram of balanced seal)	To give extra protection in the event of seal failure by reducing the pressure of the escaping product	The bush is usually made of non-ferrous material to reduce risk of sparking or seizure
AUXILIARY PACKING/back up seal	Used instead of a safety bush when conditions are so hazardous that a second or emergency safeguard must be provided	Usually consists of two turns of soft packing in an auxiliary stuffing box, or a lip seal of PTFE based material

24 Mechanical seals

SEAL PERFORMANCE

The main factors which determine the operating limits of mechanical seals are:-

1. The stability (boiling etc.) of the fluid film between the seal faces.
2. The wear life of the seal face materials.
3. The compatibility of the materials, from which the seal is made with the sealing environment.
4. Temperature of operation.

When operating correctly the sealing faces are separated by a very thin fluid film of 0.25–8 μm thick.

Rubbing contact occurs during starting and stopping, and occasionally at normal speeds. The material combinations used, therefore, need to be selected for adequate friction and wear performance, and Table 24.3 gives some general guidance.

PV values for mechanical seals are calculated in a different way from the more usual load per unit projected area times rubbing velocity.

In a mechanical seal P is taken as the pressure drop across the seal in bar and V is the mean sliding velocity at the interface in metres per second.

With unbalanced seals 100 to 140% of the sealed pressure acts on the seal faces. Balanced seals have part of the sealed pressure hydraulically relieved to reduce the force applied to the seal faces.

The materials chosen must also have good corrosion resistance, particularly since one seal face is usually carbon-graphite which generates a high electrolytic potential with most metals, when in an electrolyte.

For the components other than the seal faces, such as seal chambers, springs and shaft sleeves, the basic material is austenitic stainless steel (18/8/3) progressing as increasing chemical resistance is required to Hastelloy B (Ni 61/Co 2/Mo 28/Fe 5) and Hastelloy C (Ni 5/Co 2.5/Cr 15.5/Mo 16/W 4/Fe5). It is an advantage to coat the shaft sleeve under the sliding packing with a hard facing to reduce abrasion and corrosion.

Temperature has a major effect on the choice of packing/secondary sealing materials, as shown in Table 24.4.

Table 24.3 Typical PV limits for seal face materials corresponding to a seal life of 8000 h

Face material combination	Product	PV bar x m/s Unbalanced	Balanced
Stainless steel carbon*	Water	9	Never
	Oil	30	
Lead bronze carbon*	Water	20	Never
	Oil	35	
Stellite carbon*	Water	35	100 non-
	Oil	100	700 preferred
Tungsten carbide carbon**	Water	100	250
	Oil	150	1000
Silicon carbide carbon*	Water	150	500
	Oil	200	1500
Alumina/carbon**	Water	100	
Tungsten carbide^x tungsten carbide	Water	60	
	Oil	100	
Silicon carbide^x silicon carbide	Water	100	
	Oil	150	
Tungsten carbide^x silicon carbide	Water	100	
	Oil	300	

* metal impregnated carbon.
** resin impregnated carbon (which gives improved corrosion resistance).
^{x} for fluids containing abrasive solids.

Table 24.4a The effect of temperature on secondary sealing materials and design

Below −75°C	Vertical shafts with double seal arrangement (Seals warm gas or vapour instead of cold liquid on horizontal shaft layout)	Use synthetic rubber or PTFE packings
−75°C 30°C	Vertical or horizontal shafts with double seal arrangement	Use synthetic rubber or PTFE packings
−30°C to Ambient	Vertical or horizontal shafts with balanced seal	Use synthetic rubber or PTFE packings
Ambient to 100°C	Balanced or unbalanced seals	Use synthetic rubber or PTFE packings
100 to 250°C	Balanced or unbalanced fluoroelostomer or perfluoroelastomer rings	Use PTFE packings or glass
250 to 250°C	Balanced seal, or double seal with cooling by intermediate sealant perfluoroelastomer or graphite foil rings	Use filled PTFE wedge or packings Use PTFE packings

Table 24.4b Allowable temperature ranges for various secondary sealing materials

Fluorosilicone	−60 to 100°C
Ethylene/Propylene	−50 to 140°C
Butyl	−50 to 100°C
Neoprene	−45 to 110°C
Nitrile	−30 to 120°C
Fluoroelastomer	−25 to 180°C
Perfluoroelastomer	−10 to 250°C
PTFE (spring loaded 'O' rings)	−100 to 250°C
PTFE (glass filled wedge)	−100 to 250°C
Graphite foil	0 to 480°C

Mechanical seals 24

Table 24.5 Thermal properties of seal face materials

Material	Expansion coefficient $\times 10^6/k$	Conductivity W/mK	Diffusivity $\times 10^6 m^2/s$	Specific heat J/kg K	Maximum temperature °C continuous
Stainless steel (18/8)	16	16	4	510	600
Ni-resist	17	40	12	460	400
Alumina (99%)	8	25	7	1000	1500
Tungsten carbide (6% cobalt)	5	90	31	195	600
Silicon carbide (reaction bonded)	4	150	44	1100	1350
Silicon carbide (converted)	4	46	29	840	450
Carbon graphite					
(resin impregnated)	5	18	13	750	250
(antimony impregnated)	5	22	13	750	400
PTFE + glass	100	0.3	0.2	900	200
Stellite (cobalt based)	12	25	7	430	800
Bronze (leaded)	17	150	48	340	300

Thermal diffusivity is a measure of the rate of heat diffusion through a material (for copper the value is 112).

It equals $\dfrac{\text{thermal conductivity}}{\text{specific heat} \times \text{density}}$

High diffusivity increases the performance of a mechanical seal.
Low diffusivity can lead to damaging thermal cycling behaviour.

Table 24.6 Chemical pH tolerance of materials

Tungsten carbide (cobalt binder)	7–13	Silicon carbide (sintered alpha)	1–14
Tungsten carbide (nickel binder)	6–13	Filled PTFE	0–14
Silicon carbide (reaction bonded)	5–12	High alumina ceramic	0–14
Tungsten carbide (nickel chrome moly bonded)	2–13		

Spring arrangements

The spring arrangements which can be used depend on the size, operating speed and product sealed: guidance is given in Table 24.7.

Rubber or metal bellows can be used in place of springs.

Table 24.7 Spring arrangements suitable for various sizes and speeds

Speed rev/min	Shaft dia mm	Spring arrangement			
		Rotary		Stationary	
		Single	Multi	Single	Multi
Up to 3000	Up to 100	Yes	Yes	Yes	Yes
Up to 3000	Over 100	No	Yes	Yes	Yes
Up to 4500	Up to 75	Yes	Yes	Yes	Yes
Over 4500	Up to 100	No	No	Yes	Yes
Over 4500	Over 100	No	No	No	Yes

24 Mechanical seals

SELECTION PROCEDURE

In order to make the correct seal selection it is necessary to know:

    The product to be sealed
    Pressure to be sealed
    Shaft speed
    Shaft or sleeve dia.
    Temperature of product in the seal area

It is assumed that by knowing the product, its boiling point at the pressure to be sealed is also known.

In the procedure described here, products to be sealed are divided into aqueous products and hydrocarbons.

1. Consult Table 24.8 for Aqueous Products or Table 24.9 Hydrocarbons and note:

   The vapour pressure curve number, and possible seal face materials, seal packing and other seal component materials, for compatibility with the product, together with any special remarks for satisfactory operation.

2. Refer to the central diagram in the left-hand column of the General Seal Selection Graphs to determine for the particular shaft speed and shaft diameter an appropriate seal configuration with regard to the spring layout. Note the parameter, peripheral speed.

3. Depending whether the seal pressure is above or below 1 MN/m^2 (140 p.s.i.); refer to the Face Material Selection curves—balanced at the top of the page unbalanced at the bottom. For the particular seal pressure and peripheral speed, note the face material choices available.

4. Move to the right to the temperature and pressure curves. From the centre figure on the right, transpose from the temperature scale for the particular product corresponding to the vapour pressure curve no., to the top for a balanced seal and to the bottom for an unbalanced seal, choosing a pressure line relating to the diffusivity of the face materials (see Table 24.5). The curves then represent a stability line in terms of sealing pressure and operating temperature for particular speeds (m/sec). The area to the left of the curve is one of stable operation: the area between the curve and the saturation curve is one of instability. If the operating point falls within this latter region, either some cooling is required or the seal pressure must be changed, to bring the operating point to the left of the stability curve.

5. Compare the face material selected from corrosion and compatibility considerations from Tables 24.6, 24.8 and 24.9 with those from stability considerations as in step 4 above and decide on the appropriate combination.

6. Decide what cooling (or seal pressure change) is required; decide on the circulation system and what additional design features are required by way of quench, safety bush etc.

Selection is now complete.

Example

Water at 10 bar 3000 rev/min, 45 mm shaft dia., 175°C temperature at seal.

1. From the Aqueous Products list:
   Vapour pressure curve no. is 21
   Face material choice is stainless steel 18/8
   lead bronze, Stellite, tungsten carbide, alumina, or silicon carbide
   Seal packings are high nitrile rubber up to 100°C or ethylene propylene up to 150°C (note temperature at seal is 175°C)
   Other seal components, 17/2 stainless steel.
   No special remarks are listed.
2. From the Seal Type Selection curve either a balanced or unbalanced seal may be used (at a peripheral speed of 7 m/sec).
3. For Face Material Selection consider:
   (a) a balanced seal when tungsten carbide/carbon and silicon carbide/carbon are possibilities.
   (b) an unbalanced seal when silicon carbide/carbon is the only possible combination.
4. From the temperature and pressure curves consider
   (a) balanced seal and transposing from curve no. 21, note that at 10 bar and 7 m/sec the maximum operating temperature range is 160°C to 130°C, depending on seal face
   (b) unbalanced seal diffusivity maximum operating temperature is 70°C
5. Comparison of possible selections

Balanced seal	Cooling required from 175°C to 160°C or 130°C depending on the diffusivity of the chosen face seal.
Unbalanced seal	Cooling required from 175°C to 70°C and high diffusivity face seal.

From these possible selections, the preferred one is the balanced seal using silicon carbide running against a carbon stationary seal ring. The minimum of cooling is required and silicon carbide is chemically resistant.

Note that nitrile rubber packings are unsuitable. PTFE or graphite are suitable however. Other metal seal components are 17/2 or 18/8 stainless steel.

Mechanical seals 24

Table 24.8 Selection of materials for aqueous products

Product	vapour pressure curve no	Suitable seal ring materials for corrosion resistance*							seal packings −30 to 100°C	Other seal component materials	Remarks
		stainless steel 18/8	silicon carbide	stainless 17/2	lead bronze	Stellite	tungsten carbide	alumina			
Acetic acid	27	*	*			*			P	18/8 stainless steel	Up to 100°C only
Ammonia liquid	9	*	*			*	*	*	E	17/2 stainless steel	Use double seal indoors
Ammonium carbonate	21	*	*			*	*	*	N	17/2 stainless steel	Avoid crystallising conditions at seal
Ammonium chloride	21	*	*			*	*	*	N	18/8 stainless steel	Avoid crystallising conditions at seal. Use Monel spring
Ammonium hydroxide	21	*	*			*	*	*	N	18/8 stainless steel	
Beer	21	*	*	*	*	*	*	*	N	17/2 stainless steel	Use V packings above 100°C
Brine (calcium chloride)	21	*	*		*	*	*	*	N	18/8 stainless steel	Use Monel spring and V packings above 100°C
Brine (sodium chloride)	21	*	*		*	*	*	*	N	18/8 stainless steel	Use V packings above 100°C
Calcium carbonate	21	*	*		*	*	*	*	N	17/2 stainless steel	Avoid crystallising conditions at seal
Calcium chloride up to 100°C	21	*	*			*	*	*	N	18/8 stainless steel	Avoid crystallising conditions at seal. Use Monel spring
Carbon disulphide	20	*	*			*	*	*	V	18/8 stainless steel	Use V packings above 100°C
Citric acid up to 50% conc.	21	*	*			*			N	18/8 stainless steel	
Citric acid above 50% conc.	21								N	17/2 stainless steel	Use Hastelloy seal ring
Copper sulphate	21	*	*			*	*	*	N	18/8 stainless steel	Avoid crystallising conditions at seal Use V packings above 100°C
Dye liquors	21	*	*			*	*	*	V	18/8 stainless steel	May require P packings on occasion
Hydrogen peroxide	29		*						E	18/8 stainless steel	Use Hastelloy seal ring
Lime slurries	21	*	*			*	*	*	N	18/8 stainless steel	Use clean injection
Lye (caustic)	21	*	*			*	*	*	E	18/8 stainless steel	
Milk	21	*	*	*	*	*	*	*	N	17/2 stainless steel	
Paper stock	21	*	*		*	*	*	*	N	18/8 stainless steel	Use clean water injection with neck bush restriction
Phosphoric acid 0–20% conc.	21	*	*			*			N	18/8 stainless steel	
Phosphoric acid 20–45% conc.	21	*	*			*			E	18/8 stainless steel	
Phosphoric acid 45–100% conc.	21	*				*			P	18/8 stainless steel	
Potassium carbonate	21	*	*	*	*	*	*	*	N	17/2 stainless steel	Avoid crystallising conditions at seal
Potassium dichromate	21	*	*		*	*	*	*	E	18/8 stainless steel	Avoid crystallising conditions at seal
Potassium hydroxide up to 30% conc.	21	*	*			*	*	*	N	18/8 stainless steel	Avoid crystallising conditions at seal
Potassium hydroxide 30% to 100% conc.	21	*				*	*	*	E	18/8 stainless steel	Avoid crystallising conditions at seal
Sewage	21	*	*			*	*	*	N	18/8 stainless steel	Avoid solidification at seal. Use clean injection
Sodium bicarbonate	21	*	*	*	*	*	*	*	N	17/2 stainless steel	Avoid crystallising conditions at seal
Sodium carbonate	21	*	*			*	*	*	N	18/8 stainless steel	Avoid crystallising conditions at seal
Sodium chloride	21	*	*			*	*	*	N	18/8 stainless steel	Avoid crystallising conditions at seal. Use Monel spring
Sodium hydroxide up to 10% conc.	21	*	*			*	*	*	E	18/8 stainless steel	Avoid crystallising conditions at seal
Sodium hydroxide 10% conc. and above	21								P	Hastelloy	Use Hastelloy seal ring avoid crystallising conditions at seal
Sodium sulphate	21	*	*			*	*	*	N	18/8 stainless steel	Avoid crystallising conditions at seal
Sulphur dioxide liquid	13	*	*			*	*	*	E	18/8 stainless steel	
Water	21	*	*	*	*	*	*	*	N	17/2 stainless steel	E packings up to 150°C
Water (demineralised)	21	*	*		*	*	*	*	N	18/8 stainless steel	
Water (boiler feed)	21	*	*		*	*	*	*	N	18/8 stainless steel	
Water (sea)	21	*	*		*	*	*	*	N	18/8 stainless steel	

Seal Packing code:
E = Ethylene propylene synthetic rubber
N = High Nitrile synthetic rubber
V = Viton A fluorocarbon elastomer
P = PTFE or perfluoro-elastomer or graphite

Unless otherwise stated in remarks column PTFE can be used as the seal packing between −100°C and 250°C.
To avoid crystallising conditions at seal use the quench feature or clean solvent injection into the seal chamber through the circulation connections.

24 Mechanical seals

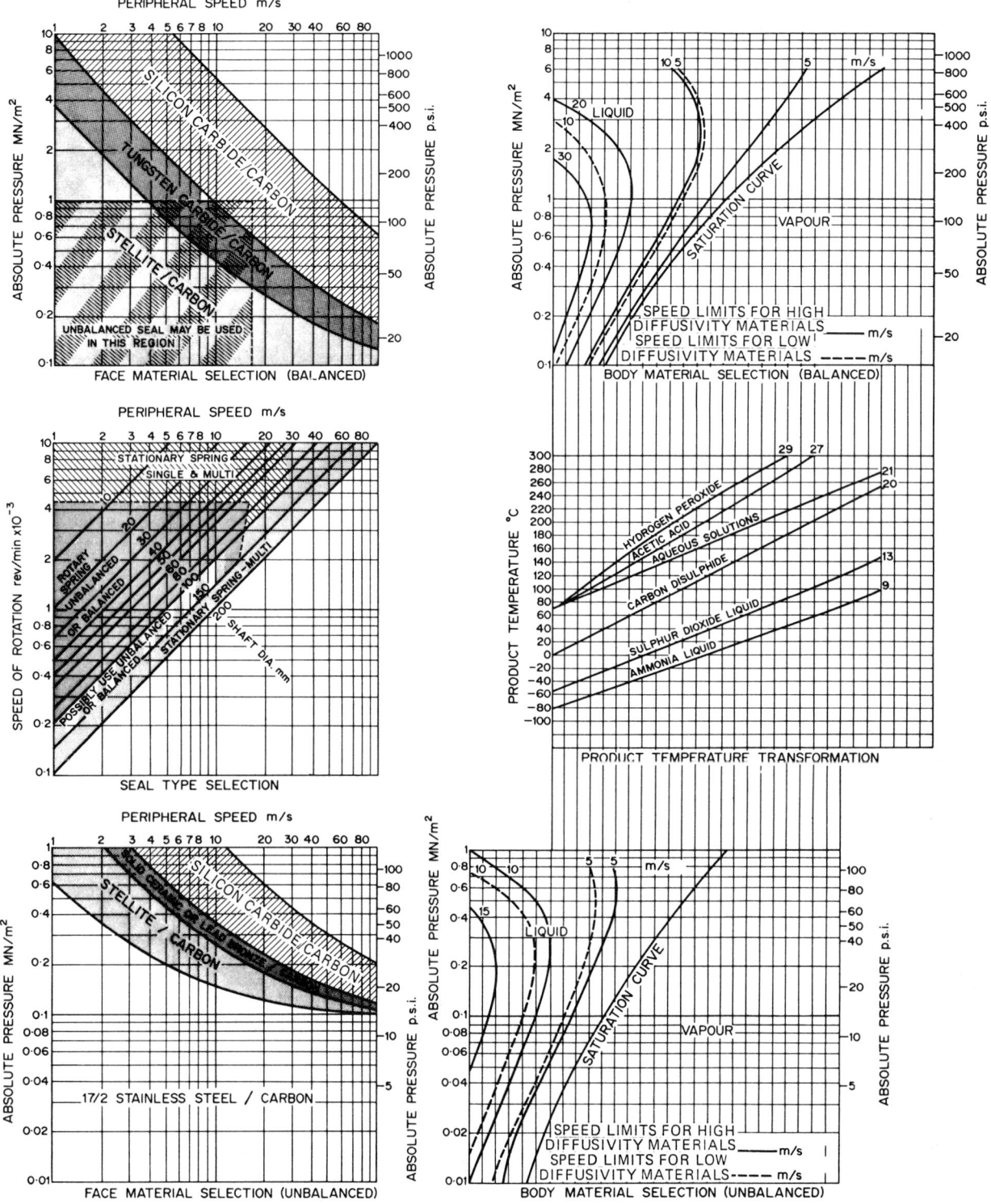

Figure 24.1 General seal selection graphs for aqueous products

Mechanical seals 24

Table 24.9 Selection of materials for hydrocarbons

Product	vapour pressure curve no	stainless steel 18/8	silicon carbide	stainless steel 17/2	lead bronze	Stellite	tungsten carbide	alumina	seal packings −30 to 100°C	Other seal component materials	Remarks
Acetone	22	*	*	*	*	*	*	*	E	17/2 stainless steel	
Arcton 9	18	*	*	*	*	*	*	*	N	17/2 stainless steel	
Benzene	24	*	*	*	*	*	*	*	P	17/2 stainless steel	
Butadiene	15	*	*	*	*	*	*	*	N	17/2 stainless steel	
Butane	16	*	*	*	*	*	*	*	N	17/2 stainless steel	
Butyl alcohol	26	*	*	*	*	*	*	*	N	17/2 stainless steel	
Butylene	15	*	*		*	*	*	*	V	18/8 stainless steel	
Cumene	30	*	*			*	*	*	V	18/8 stainless steel	
Cyclohexane	24	*	*	*	*	*	*	*	N	17/2 stainless steel	
Dimethyl ketone	22	*	*	*	*	*	*	*	E	17/2 stainless steel	
Ethane	3	*	*	*	*	*	*	*	N	17/2 stainless steel	
Ethylene	1	*	*	*	*	*	*	*	N	17/2 stainless steel	
Ethylene glycol	32	*	*	*	*	*	*	*	N	17/2 stainless steel	
Formaldehyde	13	*	*	*	*	*	*	*	N	17/2 stainless steel	
Freon 11	18	*	*	*	*	*	*	*	N	17/2 stainless steel	
Freon 12	12	*	*	*	*	*	*	*	N	17/2 stainless steel	
Freon 22	10	*	*	*	*	*	*	*	E	17/2 stainless steel	
Furfural	30	*	*	*	*	*	*	*	E	17/2 stainless steel	Use double seals above 121°C, use permanent quench below 121°C
Hexane	22	*	*	*	*	*	*	*	N	17/2 stainless steel	
Iso butane	14	*	*	*	*	*	*	*	N	17/2 stainless steel	
Iso pentane	18	*	*		*	*	*	*	N	17/2 stainless steel	
Iso propyl alcohol	24	*	*	*	*	*	*	*	E	17/2 stainless steel	
Methyl alcohol	22	*	*	*	*	*	*	*	N	17/2 stainless steel	
Methyl chloride	12	*	*	*	*	*	*	*	P	17/2 stainless steel	Use Monel spring
Methyl ethyl ketone	24	*	*	*	*	*	*	*	E	17/2 stainless steel	
Pentane	19	*	*	*	*	*	*	*	N	17/2 stainless steel	
Phenol	32	*	*		*	*	*	*	E	18/8 stainless steel	Avoid solidification at seal
Propane	10	*	*	*	*	*	*	*	N	17/2 stainless steel	
Propyl alcohol	25	*	*	*	*	*	*	*	N	17/2 stainless steel	
Propylene	9	*	*	*	*	*	*	*	N	17/2 stainless steel	
Soap solutions	21	*	*		*	*	*		N	18/8 stainless steel	Avoid solidification at seal
Styrene	29	*	*		*	*	*	*	P	18/8 stainless steel	
Toluene	26	*	*	*	*	*	*	*	P	17/2 stainless steel	
Urea (ammonia contaminated)	21	*	*			*	*	*	P	18/8 stainless steel	Avoid crystallising conditions at seal
Urea (ammonia free)	21	*	*	*		*	*	*	P	17/2 stainless steel	Avoid crystallising conditions at seal
Vinyl chloride	14	*	*			*	*	*	P	18/8 stainless steel	

Seal Packing code
- E = Ethylene propylene synthetic rubber
- N = High Nitrile synthetic rubber
- V = Viton A fluorocarbon elastomer
- P = PTFE or perfluoro-elastomer or graphite

Unless otherwise stated in remarks column PTFE can be used as the seal packing between −100°C and 250°C.
To avoid solidification at the seal on high viscosity products, heat the seal area and circulation lines 30 minutes before start-up.
To avoid crystallising conditions at seal use the quench feature or clean solvent injection into the seal chamber through the circulation connections.

24 Mechanical seals

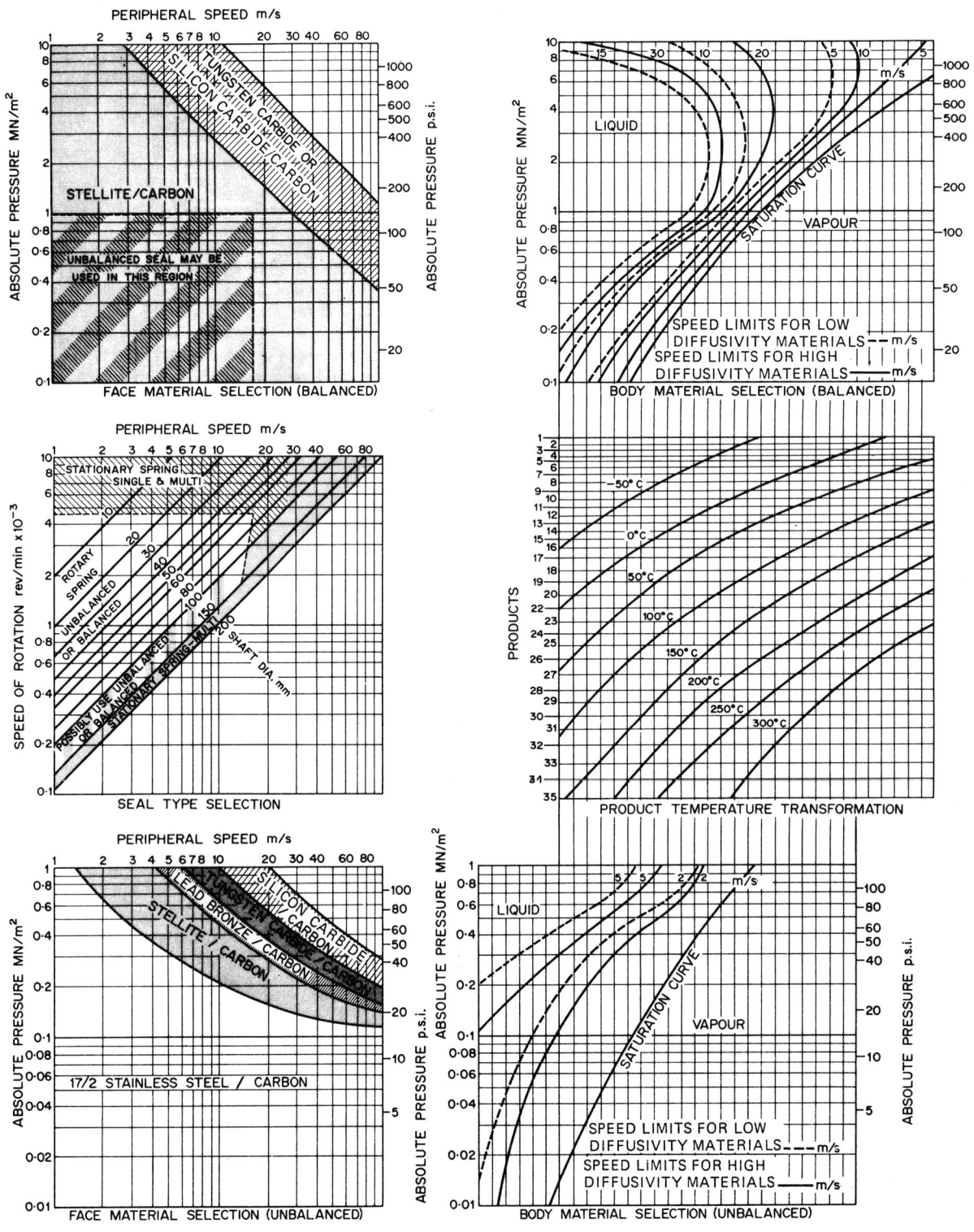

Figure 24.2 General seal selection graphs for hydrocarbons

Mechanical seals 24

Power absorption and starting torque

The coefficient of friction between the seal faces varies from a maximum dry value at zero product pressure to a minimum value in the region 0.7–1 MN/m^2 (100–140 p.s.i.). It increases imperceptibly at higher product pressures. Over the normal range of pump speeds, power absorption is proportional to speed.

The starting torque is normally about 5 times the running torque, but when a pump has been standing and the liquid film has been squeezed from between the seal faces, starting torque may be doubled. If when the pump has been standing for a long time and corrosion (chemical or electrolytic) has formed a chemical bond at the periphery of the seal rings, which has to be broken before rotation can begin, the break-out torque may increase to many times (over 5 times) the normal starting torque. Chemically inert seal faces—ceramic, glass-filled PTFE etc.—should then be used.

Power absorption and starting torque for aqueous solutions, light oils and medium hydrocarbons, use values in Figure 24.3. For light hydrocarbons use $\frac{2}{3}$ of these values. For heavy hydrocarbons use $1\frac{1}{3}$ of these values. Allow ±25% on all values.

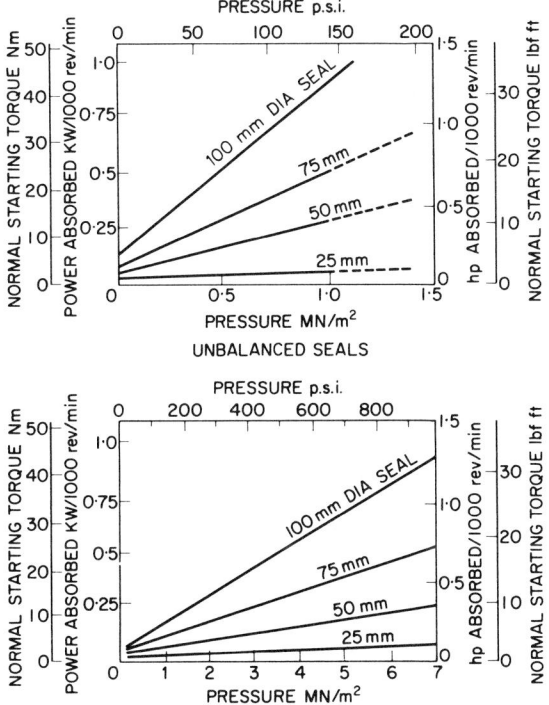

Figure 24.3 *The starting torque for various seals*

Installation and allowable malalignments

Shaft tolerances should be within ±0.05 mm (±0.002 in) and where PTFE wedge packings are used must be round within 0.01 mm (0.0005 in). Shaft surface finish should be better than 0.8 μm R_a (32 μin cla) except where PTFE wedge packings are used which require better than 0.4 μm R_a (16 μin cla) and preferably a corrosion resistant surface (Stellite) on which to slide. Where the sliding packing passes over the shaft a 10° chamfer and lead-in, free from burrs should be provided.

Seal face flatness is usually better than two or three helium light bands (0.6 or 0.9 μm) per 25 mm of working face diameter.

Seal face surface finish depends very much on the material – about 0.25 μm R_a (10 μ in cla) for carbon/graphite to about 0.05 μm R_a (2 μin cla) for hard faces.

Permissible eccentricity for O-ring and wedge fitted seals varies from 0.10 mm (0.004 in) TIR at 1000 rev/min to 0.03 mm (0.001 in) TIR at 3000 rev/min: rubber bellows fitted seals accommodate 2 to 3 times these values. Axial run out for the non-sliding seal ring is about the same order as the above but for speeds higher than 3000 rev/min a useful guide is 0.0002 mm/mm of dia (0.0002 in/inch of dia).

Axial setting is not very critical for the single spring design of seal ±2.5 mm (±0.100 in): for the multi spring design however, it is limited to about ±0.5 mm (±0.020 in).

25 Packed glands

The main applications of packed glands are for sealing the stems of valves, the shafts of rotary pumps and the plungers of reciprocating pumps. With a correct choice of gland design and packing material they can operate for extended periods with the minimum need for adjustment.

VALVE STEMS

Valve stem packings use up to 5 rings of packing material as in Figure 25.1. For high temperature/high pressure steam, moulded rings of expanded graphite foil material are commonly used. This gives low valve stem friction.

Figure 25.1 A typical valve stem packing

To reduce the risk of extrusion of the lamellar graphite during frequent valve operation, the end rings of the packing can be made from graphite/yarn filament.

Materials of this type only compress in service by a small amount and can provide a virtually maintenance free valve packing if used with live loading as shown in Figure 25.2.

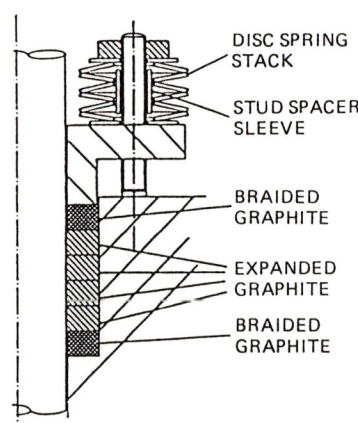

Figure 25.2 A valve stem packing using spring loading to maintain compression of the valve packing and avoid leakage

ROTARY PUMPS

Rotary pump glands commonly use up to 5 rings of packing material. For most applications up to a PV of 150 bar m/sec (sealed pressure × shaft surface speed) a simple design as in Figure 25.3 is adequate. In most pumps the pressure at the gland will be 5 bar or less and those with pressures over 10 bar will be exceptional.

At PV values over 150 bar m/sec direct water cooling or jacket cooling are usually necessary and typical arrangements are shown in Figure 25.4 and 25.5.

When pumping abrasive or toxic fluids there may be a need to provide a flushing fluid entry at the fluid end of the glands, as in Figure 25.6, or a high pressure barrier fluid which is usually injected near the centre of the gland as in Figure 25.7.

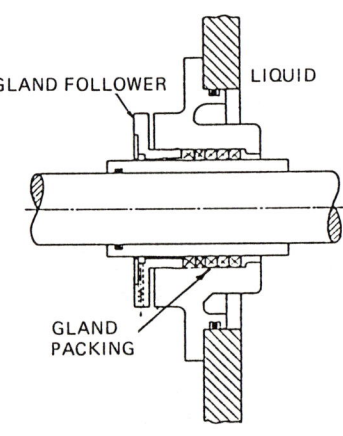

Figure 25.3 A general duty rotary pump gland

Packed glands 25

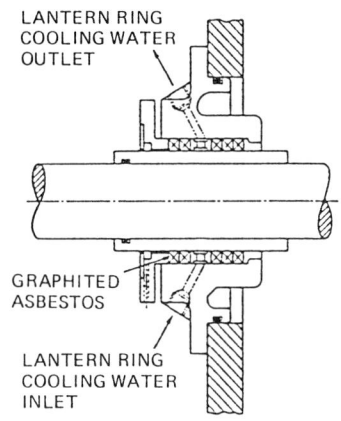

Figure 25.4 A gland packing with direct cooling via a lantern ring

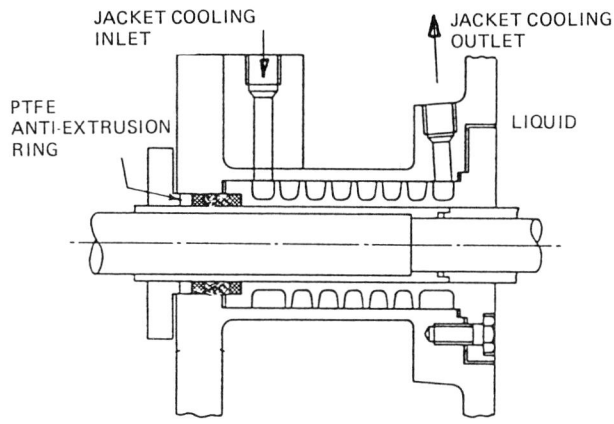

Figure 25.5 A gland packing with a cooling jacket for high temperature applications

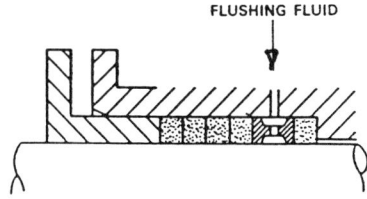

Figure 25.6 A packing gland with a flushing fluid system

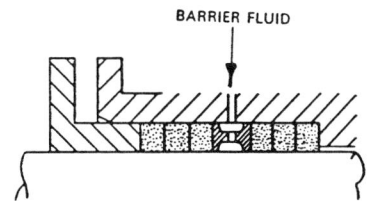

Figure 25.7 A packed gland with a barrier fluid system

RECIPROCATING PUMPS

Reciprocating pumps also use typically 5 packing rings. However due to the increased risk of extrusion of the packing due to the combination of high pressure and reciprocating movement, anti extrusion elements are usually incorporated in the gland.

Self adjusting glands can be used on reciprocating pumps but the spring loading for compression take up must act in the same direction as the fluid pressure loading, as shown in Figure 25.10.

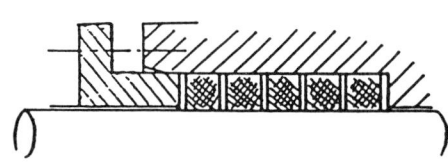

Figure 25.8 A reciprocating pump gland with PTFE anti-extrusion washers between the packing rings

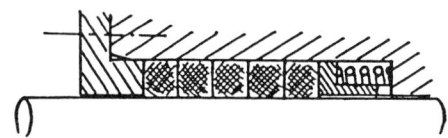

Figure 25.10 A reciprocating pump gland with internal spring loading to maintain compression of the packing

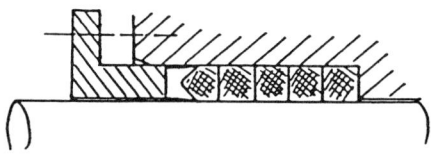

Figure 25.9 A reciprocating pump gland with an anti-extrusion moulded hard fabric lip seal

133

25 Packed glands

PACKING MATERIALS

Table 25.1 Materials for use in packed glands

Material	Maximum operating temperature °C	Special properties	Typical applications
Expanded graphite foil	550°C 2500°C in non-oxidising environments	Low friction, self lubrication, low compression set and contains no volatile constituents. Available as rings	Valve stems
Graphite/yarn filament	550°C	Available as cross plaited square section lengths. Resistant to extrusion	Valve stems
Aramid (Kevlar) fibre	250°C	Tough and abrasion resistant	Valve stems and pumps
PTFE filament	250°C	Low friction and good chemical resistance	Valve stems Pumps at surface speeds below 10 m/s
Hybrid graphite/PTFE yarn	250°C	Particularly suitable for high speed rotary shafts. Close bush clearances needed to reduce risk of extrusion. Good resistance to abrasives.	Pump shafts for speeds of the order of 25 m/s
Ramie	120°C	Good water resistance	Rotary and reciprocating water pumps

Note: Many of these packings can be provided with a central rubber core which can increase their elasticity and thus assist in maintaining the gland compression. Their application depends on the temperature and chemical resistance of the type of rubber used.

PACKING DIMENSIONS AND FITTING

Typical gland dimensions are shown in Figure 25.11 and packing sizes in Table 25.2.

Table 25.2 Typical radial housing widths in relation to shaft diameters. All dimensions in mm

All packings except expanded graphite		Expanded graphite	
Shaft diameter	Housing radial width	Shaft diameter	Housing radial width
Up to 12	3	up to 18	3
above 12 to 18	5	above 18 to 75	5
18 to 25	6.5	75 to 150	7.5
25 to 50	8	150 and above	10
50 to 90	10		
90 to 150	12.5		
150	15		

Packed glands 25

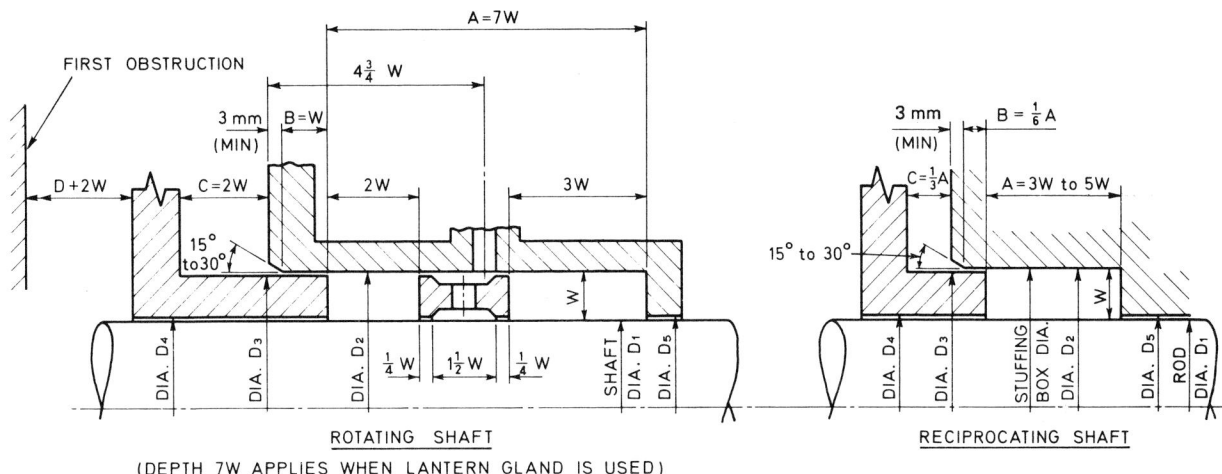

Figure 25.11 Typical gland dimensions for rotating and reciprocating shafts

Pump shafts, valve stems and reciprocating rams should have a surface finish of better than 0.4 μm Ra. Their hardness should not be less than 250 Brinell.

Rings should be cut with square butt joins and each fitted individually with joins staggered at a minimum of 90°. After applying a small degree of compression to the complete set, gland nuts must be slackened off to finger tight prior to start up. Once running, any excessive leakage can then be gradually reduced by repeated small degrees of adjustment. The major cause of packing failure is excessive compression, particularly at the initial fitting stage.

Further advice may be obtained from packing manufacturers.

26 Mechanical piston rod packings

The figure shows a typical general arrangement of a mechanical rod packing assembly. The packing (sealing) rings are free to move radially in the cups and are given an axial clearance appropriate to the materials used (see Table 26.2). The back clearance is in the range of 1 to 5 mm. ($\frac{1}{32}$ to $\frac{3}{16}$ in). The diametral clearance of the cups is chosen to prevent contact with the rod; it lies typically in the range 1 to 5 mm ($\frac{1}{32}$ to $\frac{3}{16}$ in). The sealing faces on the rings and cups are accurately ground or lapped.

The case material can be cast iron, carbon steel, stainless steel or bronze to suit the chemical conditions. It may be drilled to provide lubricant feed to the packing, to vent leakage gas or to provide water cooling.

The rings are held in contact with the rod by spring pressure; sealing action however, depends on gas forces which hold the rings radially in contact with the rod and axially against the next cup.

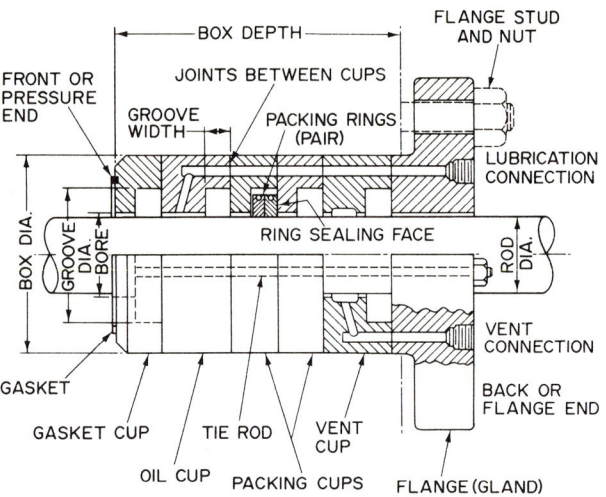

Figure 26.1 General arrangement of a typical mechanical piston rod packing assembly

SELECTION OF TYPE OF PACKING

1 Pressure breaker

Description

Three-piece ring with bore matching rod. Total circumferential clearance 0.25 mm. Garter spring to ensure contact with rod.

Applications

Used in first one or two compartments next to high pressure, when sealing pressure above 35 bar (500 p.s.i.) to reduce pressure and pressure fluctuations on sealing rings.

2 Radial cut/Tangential cut pair

Description

The radial cut ring is mounted on the high pressure side. (Two tangential cut rings can be used when there is a reversing pressure drop.) The rings are pegged to prevent the radial slots from lining up. Garter springs are fitted to ensure rod contact. Ring bores match the rod.

Applications

The standard design of segmental packing. Used for both metallic and filled PTFE packings.

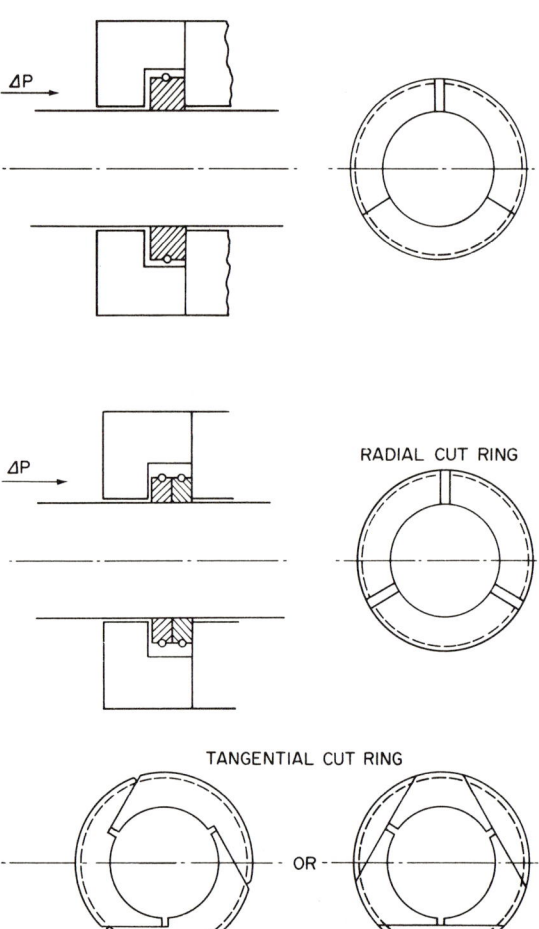

Mechanical piston rod packings 26

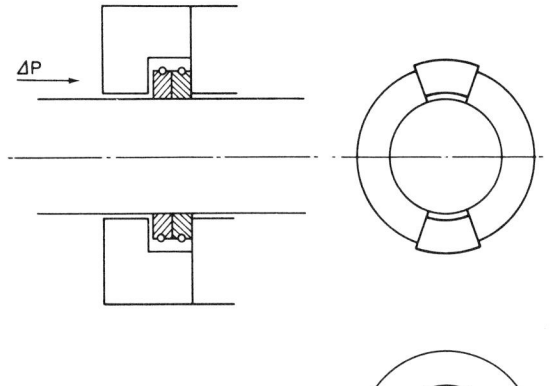

3 Unequal segment ring

Description

The rings are pegged to prevent the gaps lining up. Garter springs are fitted to ensure rod contact. The bore of the larger segment matches the rod.

Applications

Rather more robust than tangentially cut rings (2) and hence more suitable for carbon-graphite packings.

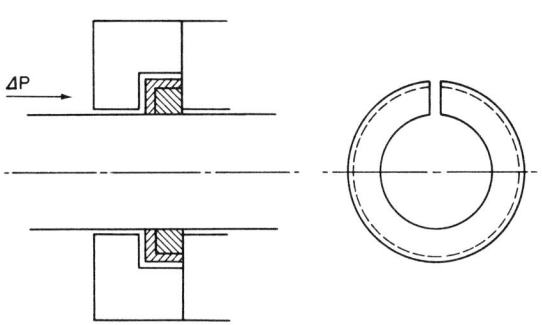

4 Contracting rod packing

Description

Cast iron L-ring with bronze or white metal inner ring or three piece packing with filled PTFE and metallic back-up ring. Contact with rod maintained by ring tension. Rings pegged to prevent the gaps from lining up. *Note*: this style of packing has to be assembled over the end of the rod.

Applications

Used for both metallic and filled PTFE packings.

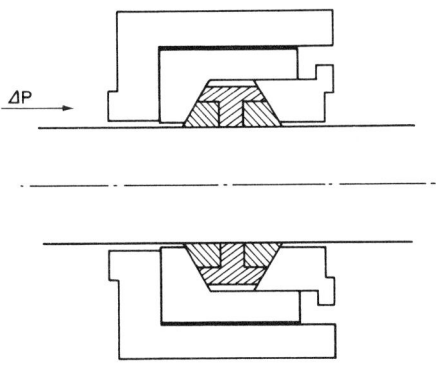

5 Cone ring

Description

Three ring seal—each ring in three segments with bore, matching rod. Cone angle ranging from 75° at pressure end to 45° at atmosphere end.

Applications

Used for both metallic and filled PTFE packings.

26 Mechanical piston rod packings

DESIGN OF PACKING ARRANGEMENT

Number of sealing rings

There is no theoretical basis for determining the number of sealing rings. Table 26.1 gives values that are typical of good practice:

Table 26.1 The number of sealing rings for various pressures

Pressure	No. of sets of sealing rings
up to 10 bar (150 p.s.i.)	3
10–20 bar (150–250 p.s.i.)	4
20–35 bar (250–500 p.s.i.)	5
35–70 bar (500–1000 p.s.i.)	6
70–150 bar (1000–2000 p.s.i.)	8
above 150 bar (2000 p.s.i.)	9–12

Notes: 1 With Type 4 packings increase number of sealing rings by 50–100%.
2 With Type 5 packings four sets of sealing rings should be adequate.
3 Above 35 bar (500 p.s.i.) use one or two pressure breakers (Type 1) in addition, on the pressure side.

Piston rods

Rod material is chosen for strength or chemical resistance. Carbon, low alloy and high chromium steels are suitable. For the harder packings (lead bronze and cast iron) hardened rods should be used; treatment can be flame or induction hardening, or nitriding. Chrome plating or high chromium steel is used for chemical resistance.

Surface finish

Metal and filled PTFE packing 0.2–0.4 μm R_a (8–16 μin cla).
Carbon/graphite and metal/graphite sinters 0.1–0.2 μm R_a (4–8 μin cla)

Dimensional tolerances

Diameter $\begin{cases} +0.05 \text{ mm } (+0.002 \text{ in}) \\ -0.05 \text{ mm } (-0.002 \text{ in}) \end{cases}$
Taper over stroked length   0.01 mm (0.0005 in)
Out-of-roundness   0.025 mm (0.001 in)

Packing materials

Table 26.2 The types of packing material and their applications

Material	Rod hardness	Axial clearance	Applications
(1) Lead-bronze	250 BHN min	0.08–0.12 mm (0.003–0.005 in)	Optimum material with high thermal conductivity and good lubricated bearing properties. Used where chemical conditions allow. Suitable for pressures up to 3000 bar (50 000 p.s.i.)
(2) Flake graphite grey cast iron	400 BHN min	0.08–0.12 mm (0.003–0.005 in)	Cheaper alternative to (1); bore may be tin coated to assist running in. Suitable up to 70 bar (1000 p.s.i.) for lubricated operation
(3) White metal (Babbitt)	not critical	0.08–0.12 mm (0.003–0.005 in)	Used where (1) and (2) not suitable because of chemical conditions. Preferred material for high chromium steel and chrome-plated rods. Max. pressure 350 bar (5000 p.s.i.). Max. temperature 120°C
(4) Filled PTFE	400 BHN min	0.4–0.5 mm (0.015–0.020 in)	Suitable for unlubricated and marginally lubricated operation as well as fully lubricated. Very good chemical resistance. Above 25 bar (400 p.s.i.) a lead bronze backing ring (0.1/0.2 mm) clear of rod should be used to give support and improved heat removal
(5) Reinforced p.f. resin	not critical	0.25–0.4 mm (0.010–0.015 in)	Used with sour hydrocarbon gases and where lubricant may be thinned by solvents in gas stream
(6) Carbon-graphite	400 BHN min	0.03–0.06 mm (0.001–0.002 in)	Used with carbon-graphite piston rings. Must be kept oil free. Suitable up to 350°C
(7) Graphite/metal sinter	250 BHN min	0.08–0.12 mm (0.003–0.005 in)	Alternative to (4) and (6)

Mechanical piston rod packings 26

FITTING AND RUNNING IN

1. Cleanliness is essential so that cups bear squarely together and to prevent scuffing or damage at start up.
2. Handle segments carefully to avoid damage during assembly.
3. Check packings float freely in cups.
4. With lubricated packings, check that plenty of oil is present before starting to run-in. Oil line must have a check valve between the lubricator and the packing. Manually fill the oil lines before starting. Use maximum lubrication feed rate during run-in.
5. If the temperature of the rod rises excessively (say above 100°C) during run-in, stop and allow to cool and then re-start run-in.
6. Run in with short no-load period.

27 Soft piston seals

SELECTION AND DESIGN

Table 27.1 Guidance on the selection of basic types

Type name	Distributor	'U'	Cup	'O' ring
External—fitted to piston, sealing in bore			(controlled compression)	
Internal—fitted in housing, sealing on piston or rod				
Simple housing design	Good	Good	Poor	Very good
Low wear rate	Very good	Good	Good	Poor
High stability (resistance to roll)	Good	Fair	Very good	Poor
Low friction	Fair	Fair	Fair	Good
Resistance to extrusion	Good	Good	Good	Fair
Availability in small sizes	Fair	Good	Poor	Very good
Availability in large sizes	Good	Fair	Good	Good
Bidirectional sealing	Single-acting only. Use in pairs back-to-back for double-acting. 'Non-return' valve action can be useful			Effective but usually used in pairs
Remarks	Do not allow heel to touch mating surface except under high pressure. Use correct fits and guided piston, etc. If seal too soft for pressure, lip may curl away from surface			Avoid parting line flash on the sealing surfaces. Unsuitable for rotational movement

Application notes

Long lips take up wear better and improve stability but increase friction. Use plastic back-up rings to reduce extrusion at high pressures. The use of a thin oil will reduce wear but may increase friction. For pneumatic assemblies use light grease which may contain colloidal graphite or MoS_2. Choose light hydraulic oil for mist lubrication.

Avoid metal-to-metal contact due to side loading or piston weight. If seals will not maintain concentricity use acetal resin, nylon, PTFE, glass fibre/PTFE or metal bearings.

To prevent mixing of unlike fluids, e.g. aeration of oil, use two seals and vent the space in between to atmosphere.

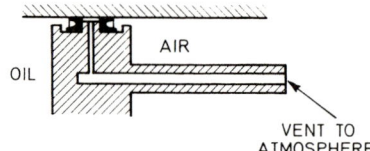

Soft piston seals 27

Table 27.2 Seals derived from basic types

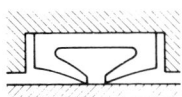 Double-acting, one-piece, narrow width, but pressure can be trapped between lips and seal may jam. Needs composite piston

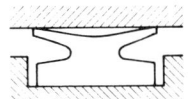 Similar, but no pressure trap and can be fitted to one-piece piston

 Derived from 'O' ring. Less tendency to roll. Improved and multiple sealing surfaces. Sealing forces reduced and parting line flash removed from working surface

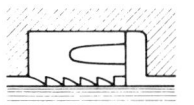 Multiple sealing lips to obviate leakage due to curl

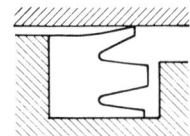 'W' section. Good for hydraulic applications and high pressures. Can be used internally or externally

Table 27.3 Special seals

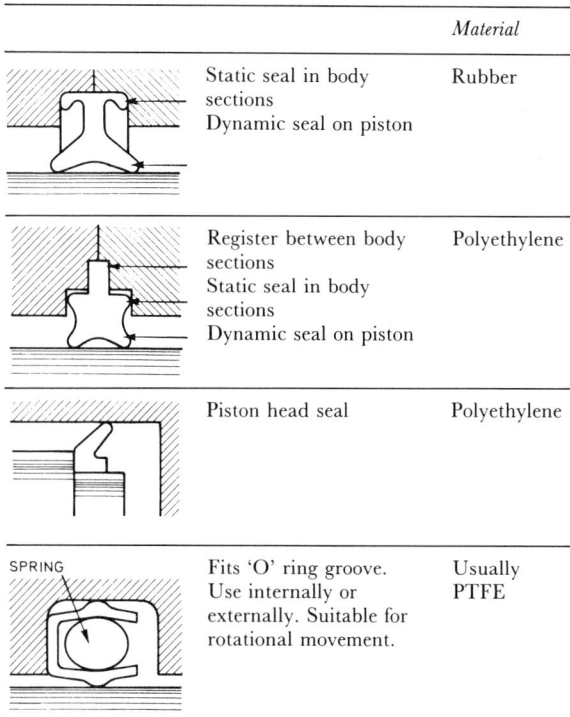

		Material
	Static seal in body sections Dynamic seal on piston	Rubber
	Register between body sections Static seal in body sections Dynamic seal on piston	Polyethylene
	Piston head seal	Polyethylene
	Fits 'O' ring groove. Use internally or externally. Suitable for rotational movement.	Usually PTFE

Table 27.4 Mating surface materials

Materials	Type	Finish		Remarks
		0.6 μm max. 0.2 to 0.4 μm (8 to 16 μin) preferred	0.2 μm max. 0.05 to 0.1 μm (2 to 4 μin) preferred	
Brass	As drawn	✓		Best untreated materials. Improve with use. High cost.
Copper	As drawn	✓		
Aluminium alloy	As drawn	✓		Liable to scuff and corrode. Low cost
	Polished		✓	
	Anodised	✓		Short life
	Hard anodised		✓	Abrasive, therefore polish
Mild steel	As drawn	✓		Corrodes. Low cost.
	Honed	✓		Corrodes readily
	Hard chrome plated		✓	Very abrasive, polish before and after plating. High cost.
	Ground		✓	Used mostly for piston rods
Stainless steel	Ground		✓	

Notes: Anodising and plating can be porous to air causing apparent seal leakage. The finish on the seal housing can be 0.8 μm. Use rust prevention treatment for mild steel in storage.

27 Soft piston seals

INSTALLATION

Table 27.5 Assembly hazards

Problem		Suggested solution
Multiple seal grooves		Fit in this order
External grooves		Bone or plastic blade – use light grease
Internal grooves		Tilt
Crossing ports		Deburr, chamfer, use assembly sleeve or temporary plug in port
Crossing threads		Use thin wall sleeve
Crossing edges and circlip grooves		Deburr or chamfer
Fitting piston assemblies to bores		A / B

Soft piston seals 27

Table 27.6 'O' ring fits

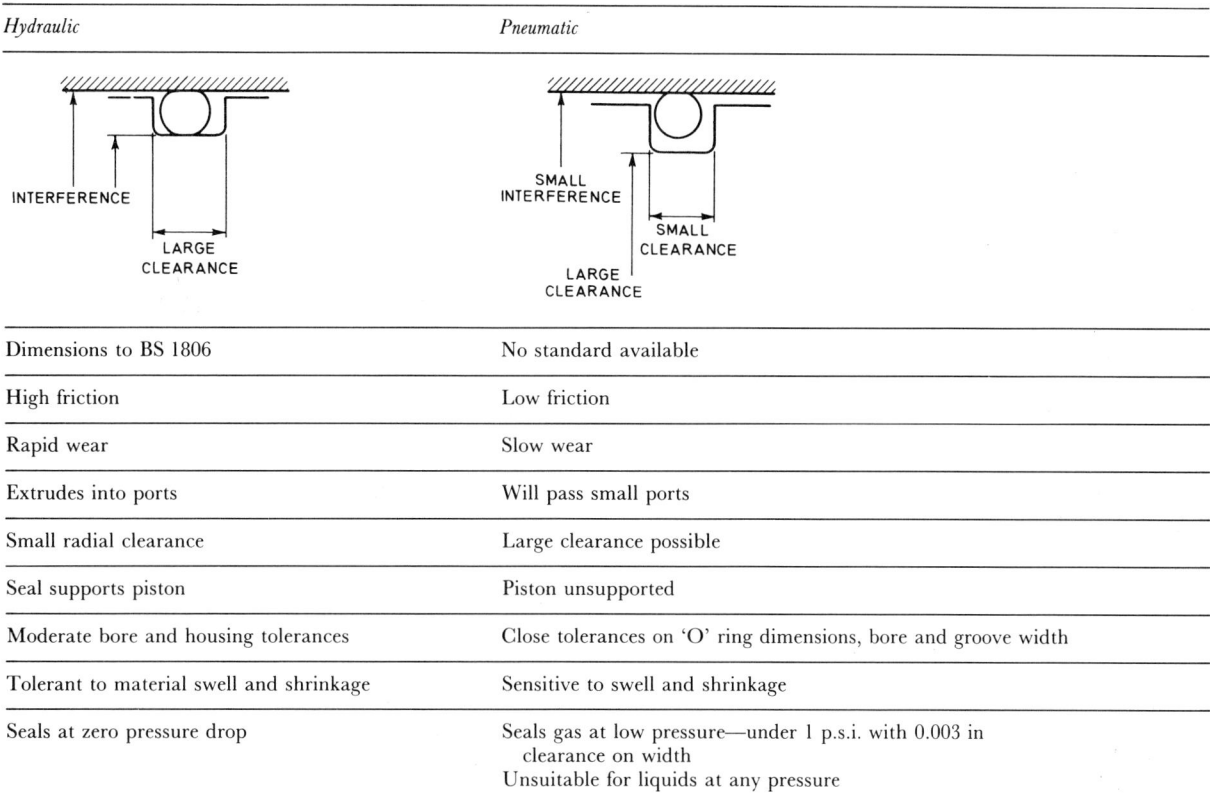

Hydraulic	Pneumatic
Dimensions to BS 1806	No standard available
High friction	Low friction
Rapid wear	Slow wear
Extrudes into ports	Will pass small ports
Small radial clearance	Large clearance possible
Seal supports piston	Piston unsupported
Moderate bore and housing tolerances	Close tolerances on 'O' ring dimensions, bore and groove width
Tolerant to material swell and shrinkage	Sensitive to swell and shrinkage
Seals at zero pressure drop	Seals gas at low pressure—under 1 p.s.i. with 0.003 in clearance on width. Unsuitable for liquids at any pressure

FAILURE

Table 27.7 Types and causes of failure

Type	Usual symptom	Cause
Channelling (fluid cutting)	Small, straight grooves across the sealing surface	Fluid leaking across seal at high velocity
Abrasive wear	Flat on 'O' ring Circumferential groove on lip seal Sharp sealing edge on lip seal	Pressure too high or abrasive mating surface
Extrusion	Surface broken Slivers of rubber	Pressure too high or too much clearance
Chemical attack	Softening or hardening—may break up	Incompatible fluid
Temperature effects	Hardening and breaking up Breaking up	Too hot and/or excess friction Too cold

Notes: Symptoms of contamination by solid particles are similar to channelling but the grooves are less regular. Uneven distribution of wear suggests eccentricity or side loading. 'O' ring rolling produces variation in shape and size of section.

Index

Air filters, 108

Ball screws, 54
Band brakes, 46, 50, 51
Bearer bands, 80
Belt drives, 1
    belt tensions, 7
    design power ratings, 4
    drive design, 1
    materials selection, 2
    multi-drive systems, 10
    pulley crowning, 9
    pulley design, 9
    pulley materials, 8
    shaft loading, 8
Bevel gears, 17, 23
Bowden cables, 75
Brakes, 46
    allowable operating conditions, 53
    areas for various duties, 49
    band brakes, 46
    disc brakes, 48
    drum brakes, 47
    materials, 52
    mating surfaces, 53
    methods of actuation, 51
    selection, 50, 51
    torque capacity, 49
Brush seals, 111
    dimensions, 114

Cams and followers:
    allowable contact stress, 61
    classification types, 59
    contact stress, 60
    design, 58
    film thickness, 61
    lubrication, 61
    modes of failure, 57
    oil & additives, 62
    surface finish, 62
    surface treatments, 62
Capstans, 70
    barrels, 71
    friction, 70
    friction coefficient, 71
    surge wheels, 71
    traction, 70
Centrifugal clutches, 39
Chevron seals, 98
Clearance seals, 99
Cone clutches, 38
Contoured disc couplings, 25, 28
Control cables, 74
    efficiency, 75
    fatigue life, 75
    load capacity, 74
    performance, 74
    pulley groove form, 75
    pulley size, 75
    selection, 74
Convoluted axial spring couplings, 26
Cylinders and liners, 94
    bore finish, 95
    design, 94
    interference fits, 95
    materials, 94, 96
    tolerances, 95

Damping devices, 76
    friction dampers, 76
    general characteristics, 76
    hydraulic dampers, 76
    performance, 77
    selection and design, 78
Disc brakes, 48, 50, 51
Disc clutch, 38
Drum brakes, 47, 50, 51
Dynamic seals, 97

Elastomeric element couplings, 26
Exclusion seals, 97, 105
Expanding band clutches, 39
Expanding ring clutches, 39

Face seals, 98
Flat belts, 2
Flexible couplings, 25
    coupling performance, 27, 28
    coupling types, 25
    effect on critical speeds, 31
    misalignments allowable, 29
    performance, 28
Friction clutches, 37
    allowable operating conditions, 44
    applications, 44
    coefficient of friction, 41
    design, 41
    design of oil-immersed clutches, 42
    duty rating, 41
    effect of temperature on wear, 45
    fitting of linings, 43
    material selection, 41
    mating surfaces, 43
    operating methods, 38, 40
    selection, 37
    types, 37
Friction dampers, 76

Gear couplings, 25, 28
    bearing loads generated, 31
    lubrication requirements, 30
    maximum misalignment, 30

Gears, 17
  AGMA, 23
  allowable stresses, 19, 20
  British standard 426 spur & helical gears, 21
  British standard 545 (bevel gears), 23
  British standard 721 (worm gears), 23
  Buckingham stress formula, 21
  choice of materials, 19
  ISO 60 spur & helical gears, 22
  Lewis formula, 21
  material combinations, 20
  non-metallic, 20
  performance, 18, 21
  torque capacity, 18
  types, 17
Gudgeon pins, 86

Helical gears, 17, 21, 22
Hydraulic dampers, 76, 77
  force-velocity characteristics, 78
Hydraulic pistons, 79

Labyrinths, 111, 114
  arrangements, 114
  leakage rates, 115
  performance, 114
Lip seals, 98
  design, 117
  extrusion clearance, 121
  friction, 121
  materials, 120
  mating surfaces, 121
  operating conditions, 118
  performance, 118
  positive-action seals, 119
  reciprocating shafts, 120
  storage and fitting, 119
  types of rubber, 119

Magnetic clutches, 40
Mechanical piston rod packings, 136
  design, 138
  fitting and running in, 139
  materials, 138
  piston rod specification, 138
  selection, 136
  typical arrangement, 136
Mechanical seals, 122
  allowable misalignments, 131
  design for aqueous products, 128
  design for hydrocarbons, 130
  effect of temperature, 124
  face materials, 125
  flush, 123
  materials for aqueous products, 127
  materials for hydrocarbons, 129
  performance, 124
  power absorption, 131
  PV limits for materials, 124
  quench, 123
  secondary sealing, 124
  selection, 126
  spring arrangements, 125
  starting torque, 131
  types, 122, 123
Metal bellows, 98
Multi-plate clutches, 38
Multiple membrane couplings, 25, 28

O rings, 143
Oil flinger rings, 109
  detail dimensions, 110
  general proportions, 109
One-way clutches, 35
  characteristics, 36
  locking needle roller, 35
  locking roller, 35
  ratchet and pawl, 35
  sprag clutch, 35
  torque and speed limitations, 36
  wrap spring, 35

Packed glands, 98, 132
  barrier fluid, 133
  cooling jackets, 133
  dimensions, 134, 135
  flushing fluid, 133
  lantern rings, 133
  materials, 134
  reciprocating pumps, 133
  rotary pumps, 132
  valve stems, 132
Piston rings, 87
  coatings, 88
  compression rings, 87
  design, 88
  dimensions, 90
  fitting stress, 89
  joints, 93
  materials, 88
  non-metallic, 92
  oil control rings, 87
  ovality, 89
  pressure distribution, 89
  ring pack arrangements, 91
  selection, 87
  side clearance, 90
Pistons, 79
  air and gas compressors, 80
  bolted crowns, 82
  compression height, 81
  design, 83
  diesel engines, 82
  dimensions, 84
  gasoline engines, 81
  gudgeon pin dimensions, 86
  hydraulic, 79
  hydraulic pumps, 80
  land widths, 85
  materials, 83
  pneumatic, 79
  ring arrangements, 83
  ring groove inserts, 82
Plain screws, 54
Planetary roller screws, 54

Quill shafts, 26

Rails, 63
   contact stress, 66
   life, 64
   vertical wear, 64
   wear, 63
Reciprocating pump glands, 133
Roller chain drives, 11
   ANSI chain drives, ratings chart, 14
   BS/DIN chain drives, ratings chart, 13
   installation and maintenance, 16
   lubrication, 14
   selection, 11
   wheel materials, 12
Rope drums, 71
Rotary pump glands, 132

Screws, 54
   installation, 56
   maximum unsupported length, 56
   mechanical efficiencies, 55
   performance, 55
   types, 54
Sealing against dirt, 105
   reciprocating parts, 107
   rotary parts, 106
   use of lip seals, 118
Sealing rings, 80
Seals, 97
   buffer fluid, 99
   'double seals', 99
   leakage rates, 104
   limits of pressure, 102
   limits of rubbing speed, 102
   multiple seals, 99
   sealing situations, 100
   selection, 97, 100
   sizes, 103
   'tandem seals', 99
   types, 97
   upper temperature limits, 101
Self-synchronising clutches, 32
   applications, 34
   design & operation, 32
   dimensions and weights, 33

   operating conditions, 34
   spacer clutches, 33
Soft piston seals, 140
   design, 140
   failure, 143
   installation, 142
   mating surface materials, 141
   selection, 140
Spur gears, 17, 21, 22
Static seals, 97
Synchronous belts, 2

Throttling bushes, 111
   leakage, 111, 112, 113
   materials, 113
Tyres, 63
   adhesion, 67, 69
   load capacity, 65
   rolling resistance, 67, 69
   tread life, 68

Valve stem packings, 132
Vee belts, 2
Vee-ribbed belts, 2
Viscoseals, 116

Wheels, 63
   adhesion, 67, 69
   load capacity, 65
   rolling resistance, 67, 69
   tread life, 68
   wear, 63
Wind-back seals, 116
Wire rope drums, 71
Wire ropes, 72
   construction, 72
   fatigue, 72, 73
   industrial applications, 72
   loading, 73
   mining applications, 73
   performance, 72, 73
   selection, 72
Worm gears, 18, 23